# Columbia University

## Contributions to Education

### Teachers College Series

No. 805

AMS PRESS
NEW YORK

# A Study of the Problems in Teaching the Slide Rule

*By* CARL N. SHUSTER

SUBMITTED IN PARTIAL FULFILLMENT OF THE REQUIREMENTS FOR THE DEGREE OF DOCTOR OF PHILOSOPHY IN THE FACULTY OF PHILOSOPHY, COLUMBIA UNIVERSITY

*Published with the Approval of*
*Professor Clifford B. Upton, Sponsor*

Bureau of Publications
Teachers College, Columbia University
New York · 1940

**Library of Congress Cataloging in Publication Data**

Shuster, Carl Nathaniel, 1890-
A study of the problems in teaching the slide rule.

Reprint of the 1940 ed., issued in series: Teachers College, Columbia University. Contributions to education, no. 805.
Originally presented as the author's thesis, Columbia
Bibliography: p.
1. Slide-rule. I. Title. II. Series: Columbia University. Teachers College. Contributions to education, no. 805.
QA73.S53 1972 510'.28 72-177784
ISBN 0-404-55805-4

Reprinted by Special Arrangement with Teachers College Press, New York, New York

From the edition of 1940, New York
First AMS edition published in 1972
Manufactured in the United States

AMS PRESS, INC.
NEW YORK, N. Y. 10003

## ACKNOWLEDGMENTS

An important prerequisite for a study of this kind is a large number of loyal friends. The author has been exceptionally fortunate in the friends who have helped him in the various phases of this undertaking, and wishes to express his gratitude and appreciation to them.

The author is especially indebted to Professor Clifford B. Upton for the generous amount of time that he devoted to the constructive criticism of this dissertation, and to Professors William D. Reeve, Henry A. Ruger, Helen M. Walker, and Robert von Nardroff for advice and suggestions during the progress of this investigation. He also acknowledges his obligations to Dr. A. Day Bradley and Dr. Fred L. Bedford for administering certain tests required in this study, to Miss Anna C. Paxton for assistance in trying out new techniques, and to Professor Charles R. Rounds, Dr. Aaron Bakst, and Mr. Robert B. Macdougall for technical help in the preparation of the manuscript. He also thanks President Roscoe L. West of the New Jersey State Teachers College at Trenton for his constant encouragement and advice throughout this investigation.

Carl N. Shuster

# CONTENTS


| CHAPTER | | PAGE |
|---|---|---|
| I. | INTRODUCTION | 1 |
| II. | MECHANICAL COMPUTATION | 3 |
| III. | A BRIEF HISTORY OF DECIMAL FRACTIONS | 8 |
| IV. | APPROXIMATE COMPUTATION AND THE SLIDE RULE | 15 |
| V. | ACCURACY AND THE SLIDE RULE | 26 |
| VI. | COMMON ERRORS MADE IN SLIDE RULE COMPUTATION | 37 |
| VII. | METHODS OF PREVENTING SLIDE RULE ERRORS | 43 |
| VIII. | TESTING THE EFFICIENCY OF THE METHODS OF PREVENTING SLIDE RULE ERRORS | 66 |
| IX. | SUPPLEMENTARY TESTS | 83 |
| X. | GENERAL SUMMARY | 93 |
| BIBLIOGRAPHY | | 96 |
| APPENDIX: INTRODUCTION TO THE TEACHING OF THE SLIDE RULE | | 99 |

# CHAPTER I

## INTRODUCTION

THE slide rule has been taught in many secondary schools of this country for the past twenty-five years. Its present use in these schools, however, is not as widespread as might be expected. Furthermore, many students who have been taught to use the slide rule do not seem to appreciate its value and make relatively little use of it after the period of instruction is over. There are also teachers of mathematics and science who claim that slide rule results are not sufficiently accurate for practical purposes. Much of the teaching of the slide rule has been poorly done. Many teachers seem to have little knowledge of the more frequent types of errors made in slide rule computation and of the means of preventing them.

In view of these circumstances, it is the aim of the writer to make a critical study of the following questions relating to the teaching of the slide rule:

(1) What are the chief circumstances or influences that have operated to prevent a wider use of the slide rule? Can these be overcome?

(2) Is the slide rule sufficiently accurate to be of any real value in such a scientific age as this? Can it be used in business?

(3) What degree of accuracy is expected in modern business, industry, and science in connection with measurement and computation?

(4) What training in computation and measurement should precede actual instruction in the use of the slide rule?

(5) What are the common errors made by students who are learning to use the slide rule? Can these errors be prevented, and if so, how?

(6) Are there special methods for teaching the slide rule that seem to increase interest and efficient learning, and if so, what are these methods?

The author has had twenty-five years of experience in teaching slide rule computation; he has also used the slide rule widely in engineering and in commercial work. For the past twelve years he has experimented with various techniques in the teaching of the slide rule and has drawn on the experience of many others who have been interested in slide rule computation.

In connection with this study, the author has made the following investigations:

(1) A critical study of the practice with respect to decimal fractions and approximate computation in a large number of elementary and high school textbooks and teachers' manuals in mathematics.

(2) A study of various state and national laws to find the tolerance in weighing and measuring allowed by law.

(3) A study of scientific literature and textbooks to determine the number of significant figures required in scientific work.

(4) A study of business practices based on the experience of the author as a senior cost accountant in the Federal Civil Service.

(5) A survey of the history of the slide rule.

(6) A study of approximate computation in its relation to slide rule computation.

(7) An analysis of slide rule manuals to determine the methods employed by various authors in presenting the slide rule.

(8) An experimental investigation to determine the common slide rule errors.

(9) Experiments to discover methods to prevent the common slide rule errors.

(10) An experiment to test the efficiency of the method worked out in (9) above.

(11) A statistical comparison of two groups of students taught by different methods.

## CHAPTER II

## MECHANICAL COMPUTATION

Computation was difficult with the early number systems. It is not strange, therefore, to find that a number of devices to aid computation were invented by the ancients. These devices are interesting, and many of them, with certain modifications, are still in use. Various tables, such as multiplication tables and tables of squares and cubes, were among the first aids to computation. These tables were used by the Babylonians more than 4,000 years ago,[1] and similar tables may still be found, along with many others, in mathematical textbooks and in the handbooks used by bankers, actuaries, and engineers. The most important of the devices for mechanical computation was the abacus.[2] The abacus had a large variety of forms, ranging from the sand table or counting board to other forms having beads on wires. The Japanese soroban, the Chinese suanpan, and the Russian s'choty (various modern forms of the abacus) are still used, and in the hands of a skilled operator are very satisfactory computing instruments. Most of our meters for measuring gas, electricity, water, and distance (with the automobile) are mechanical adaptations of the abacus, and almost all of our rather complex computing machines, so widely used today, are mechanical abaci.

Compasses were used by the early Greeks and other peoples for certain types of computation. It is possible to add, subtract, divide, multiply, extract square roots, and work problems in proportion with compasses. Proportional compasses[3] were used by the Romans at least 2,000 years ago, and are still to be found in the kit of a

[1] David Eugene Smith, *History of Mathematics*, Vol. I, p. 40. Ginn and Co., Boston, 1923.

[2] *Ibid.*, Vol. II, pp. 156-194.

[3] Vera Sanford, *A Short History of Mathematics*, p. 338. Houghton Mifflin and Co., Boston, 1930.

well-equipped modern engineer. The sector compass, invented by Galileo around 1600,[4] was used for a wide range of computation. In fact, many of the problems that are now solved by the slide rule could be solved by using either of these instruments. The sector compass was less convenient and less accurate than the slide rule but was popular for at least two hundred years, in spite of the fact that slide rules were invented about the same time.

Many of the early surveying instruments, such as the plane table and quadrant, permit a graphic solution of triangles and to this extent may be regarded as computing instruments. One of the popular aids to computation was "Napier's Bones" or "Rods." These Rods were invented by Napier about the year 1620.[5]

In recent years various planimeters and mechanical integrators have assumed an important role in scientific and engineering calculations. Planimeters are of special value in finding the areas of very irregular figures.

The slide rule differs from all the other mechanical computing instruments in that it is based on logarithms. Napier published his work on logarithms in 1614. Six years later Edmund Gunter[6] made a logarithmic scale and used it for mechanical computation. To compute with his scale Gunter used a pair of dividers instead of sliding a scale along its duplicate. The idea of making and using a logarithmic scale for mechanical computation at a time when graphic devices were not at all common was truly remarkable. It is also certain that almost the first person to lay two duplicate scales alongside of each other would see the advantage of employing two scales in place of dividers for adding and subtracting segments. While Gunter claimed to have invented the double scale, it seems probable that Oughtred used the double scale first in 1621, or one year after Gunter made and used his logarithmic scale. It is for this reason that Oughtred is commonly credited with the invention of the slide rule. Since all the basic ideas of the slide rule originated with Gunter, and Oughtred only improved on the method of manipulating the scales, as dozens of others have in the three hundred

[4] Sanford, *op. cit.*, pp. 340-343.
[5] *Ibid.*, pp. 339.
[6] *Ibid.*, pp. 344-347.

years that followed, Gunter should be accepted as the inventor of the slide rule. It should be remembered that Oughtred simply used two of Gunter's scales placed side by side like two rulers. He did not call it a slide rule. It was thirty years before the rule assumed anything like its present form, and nearly one hundred and fifty years before the runner or indicator was made a fixed part of it. Many people believe that Oughtred invented the slide rule as it exists today. On the contrary, there has been a constant evolution of the slide rule extending over its history of more than three hundred years. For a detailed history of the development of the slide rule, the reader is referred to Cajori.[7]

There is at present an almost endless variety of slide rules. One American firm is selling twenty-six different types. Special slide rules are made for a large number of purposes. Among these special slide rules may be mentioned the chemical, electrical, power computing, friction head, urea index, stadia, sewer, and merchant's slide rules.

The first man to compute with decimal fractions was Christian Rudolff (1530)[8] but he did not seem to comprehend their value as fully as did Stevin (1585).[9] Stevin appreciated the convenience of decimal fractions but had a poor symbolism. It was not until 1600 that the comma was used as a decimal point and some time later that the period was used.

It is rather remarkable that decimal fractions were not invented until nearly 1,000 years after the introduction of the symbol for zero and the discovery of the principle of *place value*. One naturally would expect that decimals would have followed within a few years after the development of the principle of place value, but such was not the case. Logarithms, however, were invented soon after decimal fractions, which were necessary to make logarithms practical (Napier 1614). It is even more remarkable that Gunter should have followed so soon with the invention of the slide rule (1620). The slide rule depends entirely on logarithms and decimal fractions.

[7] F. Cajori, *A History of the Logarithmic Slide Rule and Allied Instruments*. Engineering Publishing Co., New York, 1909.

[8] D. E. Smith, *op. cit.*, Vol. II, pp. 240-241.

[9] *Ibid.*

The slide rule became quite popular for a time in England. Even such men as Samuel Pepys[10] found it "very pretty for all questions of arithmetick."

However, the early slide rules were often poorly made. They were engraved by hand on metal, wood, or bone, and ranged in length from about six inches to six feet. These first slide rules were without indicators and they lacked many of the other features of the modern rules. It is certain that any early slide rule which was carefully made was very expensive in comparison with the slide rules of today.

Decimal fractions were not in general use in the seventeenth century, and the slide rule is of little value without decimal fractions. The greatest obstacle, however, to the use of the slide rule was the fact that the computers of that period had almost no concept of approximate computation or accuracy of measurement and little respect for three-figure answers. A perusal of the arithmetics of the time shows that the answers to the problems contain many more supposedly significant figures than could possibly have resulted from the measurements given in the problems.

After the first phase of popularity, the use of the slide rule was largely confined to specialists. Very little interest was shown in the slide rule in France, Germany, and other Continental countries in the century following its invention. However, after the metric system and extended use of decimals became common, more interest was shown in France than in England. After the invention of an accurate dividing engine (1821) that would graduate a number of rules at the same time with precision far greater than that of the old hand-engraved rules, the main objections to the slide rule, namely, lack of accuracy, lack of standardization, and high cost, were removed. In 1850 Lieutenant Mannheim of the French army invented the rule that bears his name. This became the standard rule in France and slowly increased in popularity in England and the United States.

As already pointed out, the slide rule was invented in England and practically all of the early improvements were made in that country. However, in Ball's *History of Mathematics* published in

[10] Sanford, *op. cit.*, pp. 343-352.

1888,[11] no mention is made of the slide rule. Ball gives considerable space, however, to Oughtred and mentions Gunter in connection with other topics. Ball's edition of 1893 likewise does not mention the slide rule. It is interesting, too, to note that the slide rule is not mentioned in Fink's *History of Mathematics*[12] or in Cajori's *History of Elementary Mathematics*.[13] By 1909, however, Cajori considered the slide rule sufficiently important to warrant a special history which he wrote at that time.[14]

The first college in the United States to make the slide rule a required part of its work in engineering was Washington University in St. Louis, Missouri (about 1880).[15] By 1890 a number of schools were beginning to use the slide rule and the engineering profession was becoming acquainted with it. The high cost of slide rules at this time hampered their use in schools. As early as 1903, Professor Upton, who was teaching the slide rule at Teachers College, Columbia University, wrote to several of the leading manufacturers saying that if they could place a fairly good rule on the market at a reasonable price, he believed it would be possible to introduce the slide rule into the high schools of the country. It was a long time before there were any really satisfactory cheap rules, but there are now a number of them on the market, selling at prices ranging from $.25 to $1.25, and the slide rule is being rapidly introduced into the high schools.

The two highest hurdles for the slide rule teacher of today are the same hurdles that existed three hundred years ago. These are: lack of experience with approximate computation, and lack of ability to use decimal fractions. In order to overcome these difficulties some changes should be made in our teaching of arithmetic. These changes not only will make it easier to teach the slide rule but will also greatly improve the subject matter of arithmetic, and of the mathematics and science of the high school and college as well.

[11] W. W. R. Ball, *A Short Account of the History of Mathematics*. London, 1888.

[12] E. Fink, *A Brief History of Mathematics*. Open Court Publishing Co., Chicago, 1890.

[13] F. Cajori, *A History of Elementary Mathematics*. Macmillan Co., New York, 1896.

[14] F. Cajori, *A History of the Logarithmic Slide Rule*. Engineering Publishing Co., New York, 1909.

[15] J. E. Thompson, *A Manual of the Slide Rule*, p. 17. New York, 1930.

# CHAPTER III

## A BRIEF HISTORY OF DECIMAL FRACTIONS

SINCE the slide rule is based on the decimal system of notation, it is necessary for slide rule students to have a thorough knowledge of decimal fractions and of all phases of computation with them. Relatively few students in college classes possess a thorough working knowledge of this phase of arithmetic, hence it is necessary to spend some time in reviewing decimal fractions before attempting to teach the slide rule. Students of the slide rule should also be familiar with the principles of approximate computation and should appreciate the value of numerical results having three-figure and four-figure accuracy. This chapter will present a brief history of the development of decimal fractions and will explain the rapid increase in the use of decimals that has come about in the past thirty-five years.

Our number system is one of the greatest achievements in the history of civilization. It is not the product of one mind or age, but the climax of the efforts of hundreds of the world's best thinkers. It was developed by a slow evolutionary process extending over thousands of years. The various phases of this evolution are fully treated in the histories of mathematics. Every student of mathematics should study these histories to see how the early number systems were developed and how and why these systems were eventually displaced by the much more satisfactory place-value number system.

With the early peoples the evolution of the fraction concept was attended by many difficulties. The old number systems were far from satisfactory for work with fractions, and many interesting methods were employed to simplify computation with fractions. The Egyptians restricted their common fractions to those having unity for a numerator. The Greeks made extensive use of sexagesimal fractions, while the Romans avoided fractions, whenever possible, by subdividing their measuring units.

From the historical standpoint decimals developed very slowly and even today decimals are not as widely used as they undoubtedly will be during the next half century. It is difficult to state just when decimals were invented, but it is probable that Rudolff (1530) and Stevin (1585) are entitled to most of the credit for bringing decimals to the attention of the world. Both Rudolff and Stevin had awkward ways of writing decimals and it was not until some time later that the comma or the dot was used to serve as a decimal point. Even today the methods of writing decimals are not uniform throughout the world. In America $23\frac{4}{10}$ is written 23.4, the period being used for the decimal point; in England it is written 23·4, the point being raised above the line of writing; and on the Continent it is written 23,4, where the comma serves as the decimal point.[1]

After the invention of decimals their use was very limited for about two hundred years, being confined largely to logarithmic tables and similar tabulated data. In 1786 decimals gained in importance when the United States abandoned the English monetary system and adopted a decimal system of coinage in its place. In 1799 the decimal movement was further promoted when the French Government adopted the metric system, though this system did not come into general use in France until 1840, when it was made obligatory.[2]

In 1866 the metric system was made legal in the United States and at that time the meter was adopted as the official standard of length in this country, the yard being defined in terms of the meter.[3] Though the metric system was made legal at that time, its actual use today in this country is confined, in general, to scientific work. From time to time attempts have been made to persuade the United States Congress to pass a bill making the metric system obligatory in this country for all kinds of measuring, but none of these attempts has succeeded. Similar attempts have been made in Great Britain but without success. If the use of the metric system should

[1] D. E. Smith, *History of Mathematics*, Vol. II, pp. 246–247. Ginn and Co., Boston 1925. [2] Sanford, *op. cit.*, p. 365.

[3] L. C. Karpinski, *History of Arithmetic*, p. 167. Rand McNally Co., Chicago, 1925.

become obligatory in these two countries, it is evident that there would soon be a widely increased use of decimals.

The slow progress of decimals during the past three hundred years has naturally been accompanied by a correspondingly slow acceptance of the slide rule as a means of computation since the slide rule is primarily a decimal instrument. Another element which has still further prevented the use of the slide rule is the fact that such decimals as were taught were presented so impractically in the school textbooks of the past century that it was impossible to use the slide rule in connection with the decimals even if one had been inclined to do so. Illustrations of the absurd way in which decimals have been presented in these textbooks are shown below, these problems being taken from texts published between 1781 and 1906.

1. What is the cube root of 7612181.7612? (Dilworth, 1781)[4]
2. What is the square root of 964.5192360241? (Adams, 1807)[5]
3. Multiply 31.00467 by 10.03962. (Davies, 1842)[6]
   (Ans. 311.2751050254)
4. Find the square root of 48303584.206084. (Vodges, 1845)[7]
5. (a) From $5.\dot{1}234\dot{5}$ take $2.3\dot{5}2345\dot{6}$.
   (Ans. $2.7\dot{7}11055821666927777988888899999\dot{4}$)
   (b) $100 \times 1.06^{30}$
   (Ans. 574.34911728132501162631063332310802645846357252196069357387776)
   (c) Divide 121932631112 by 1234.56789.
   (d) Divide 1191.0994835 by 3.812345.
   (e) Divide 4327.56284563 by 873.469. (Greenleaf, 1860)[8]
6. (a) Multiply 36.74637 by 127.0463.
   (Ans. 4668.490346931)
   (b) Divide 754.347385 by 61.34775. (Davies, 1866)[9]

[4] Thomas Dilworth, *Schoolmaster's Assistant*, p. 158. R. Aitken, Philadelphia, 1781.

[5] Daniel Adams, *Scholars Arithmetic*, p. 164. John Prentis, Keene, N. H., 1807.

[6] Charles Davies, *Arithmetic*, p. 173. A. S. Barnes and Co., Philadelphia, 1842.

[7] William Vodges, *United States Arithmetic*, p. 203. Caleb Bartlett, Philadelphia, 1845.

[8] Benjamin Greenleaf, *National Arithmetic*, pp. 199-226. Sower Barnes and Co., Philadelphia, 1860.

[9] Charles Davies, *University Arithmetic*, pp. 179 (a), 185 (b). A. S. Barnes and Co., N. Y., 1866.

7. Reduce .89375 gallon to quarts, pints, and gills. (Thompson, 1876) [10]
8. (a) From 2½ take 2½ thousandths and 2½ billionths.
   (Ans. 2.4974999975)
   (b) Reduce to a common fraction .194805 $\dot{1}5/75$.
   (c) Multiply 1.08096 by 3.5702. (Wentworth, 1898) [11]

All the above examples in decimals are wholly artificial and impractical. The numbers in these examples represent a degree of refinement in measurement that could not possibly be obtained by any measuring instrument available in those days or even today. It is very evident that the slide rule could not be used with examples containing such absurd data, since the ordinary slide rule is limited to three-figure and four-figure accuracy.

Equally impractical and uninteresting were the problems on the metric system found in these old textbooks. The following are typical examples:

1. (a) In 60.25 hectares, how many acres?
   (Ans. 148.8775 acres)
   (b) Reduce 137.75 kilos to pounds. (Thompson, 1876) [12]
   (Ans. 303.68365 lbs.)
2. In 10 lb. 4 oz. Troy, how many Kilograms? (Robinson, 1874) [13]
   (Ans. 3.82575 Kg.)
3. What is the cost of 3 kilares, 7 hektares, 6 deciares of land at $275.25 a hektare? (Brooks, 1906) [14]
   (Ans. $10,185.90)

To be obliged to work examples like those just given "cured" several generations of American pupils of any desire to see the metric system adopted in this country. Fortunately the school arithmetics written during the past ten years have eliminated such impractical problems in decimals and the metric system. Hence the present-day teaching of these topics is, in general, in terms of problems of the type that can be worked on a slide rule.

[10] J. B. Thompson, *New Practical Arithmetic*, p. 205. Clark and Maynard, N. Y.' 1876.

[11] G. A. Wentworth, *Advanced Arithmetic*, pp. 29-55. Ginn, Boston, 1898.

[12] Thompson, *op. cit.*, p. 215.

[13] D. W. Fish, *Robinson's Progressive Practical Arithmetic*, p. 487. Ivison, Blakeman, Taylor and Co., New York, 1874.

[14] Edward Brooks, *Normal Standard Arithmetic*, p. 407. Christopher Sower Co., Philadelphia, 1906.

Though decimals have not had the rapid spread that would have come if the metric system had been made obligatory in the United States and Great Britain, there has been, however, during the past thirty-five years another development of somewhat smaller magnitude that has extended the use of decimals to a point far beyond anything that existed previously. This new expansion in the use of decimals has been coincident with, and dependent upon, the rapid growth of modern industry during the same period. It is a well-known fact that our automobile factories, airplane factories, radio plants, and other industries have need for more refined measurements today than ever before, and to get this additional refinement it has become necessary to divide our *own units of measurement*, such as the inch, into *hundredths*, *thousandths*, and even *hundred-thousandths*, instead of *eighths*, *sixteenths*, and *thirty-seconds* as formerly. Similar developments have brought about the decimal divisions of other important units, such as the foot, mile, pound, hour, second, and degree.

The decimal division of the foot is met in modern surveying where the tapes are divided into *tenths* and *hundredths* of a foot and the leveling rods are divided into *thousandths* of a foot; the decimal division of the mile appears on the automobile speedometer and in automobile road books where the mile is divided into *tenths*; the decimal division of the pound is found on dairy scales and on some school scales, the pound being divided into *tenths*; the decimal division of the hour, minute, and second has been used on many time-clocks and stop-watches, while the electric timing of races has made the *hundredth* of a second a common measure. The degree is divided decimally on the fever thermometer, in angle measure, and in modern trigonometric tables. Thus we see that the United States and Great Britain have come to have a metric system of their own, based on the old English units of measure rather than on the metric units, such as the meter and the gram. It is clear that this decimal division of our own units has rapidly increased the use of decimals and, at the same time, has opened the door for a much wider use of the slide rule.

Concurrent with the somewhat technical use of decimal fractions has come a greatly increased use of decimals in newspapers and magazines. The following quotations illustrate this new practice:

"The new golf ball 1.68 inches in diameter and weighing 1.55 ounces replaces a ball 1.62 inches in diameter and weighing 1.62 ounces."

"Farmers who failed averaged 7.3 years in agriculture."

"Corn 31.1 bu., wheat 16.9 bu., oats 34.5 bu., potatoes 143.0 bu., and hay 1.76 tons per acre."

"The new Cascade tunnel is 7.79 miles long, while the St. Gothard is 9.26 miles long."

"Kaye Don broke the world's speedboat record by averaging 119.81 miles an hour."

"This model refrigerator has 26.36 sq. ft. shelf area, food storage capacity 14.35 cu. ft., and ice freezing capacity 28.50 lbs."

"Exposure—0.001 sec." "Rainfall—3.17 in."

Even farm papers today are filled with decimal fractions. A copy of the *Country Gentleman* (March, 1936) selected at random contained 179 decimal fractions, excluding United States money. Among the decimal fractions were the following:

| | |
|---|---|
| 2.25 gal. | 11.4% water |
| 1.75 grams | 5.4 test (milk) |
| 8.75 in. | 24.6 drawbar horsepower |
| 49.9 lb. | 339.5 million bushels |
| 2.9 M.P.H. | 99.7 calories per ounce |

Other decimal fractions found in recent farm publications are given below:

| | |
|---|---|
| 2.8 mm. | 14.6 months |
| 24.6 oz. | 5.67 cts. per K.W.H. |
| 87.8 acres | 11.378 miles per gallon |
| 308.6 gal. | water flow 3.75 ft. per sec. |
| 0.000025 in. | 27.85 points (egg contest) |

The most recent champion corn husker husked 31.142 bushels of corn (by weight) in 80 minutes. Farm papers certainly expect farmers to be familiar with decimal fractions.

A recent trend, emphasizing both decimals and the metric system, is found in the practice of stating the weights of packaged food products by giving both the English unit and its metric equivalent.

For example, a package may be labeled "Net weight: 7.5 oz. or 213 grams."

The foot rule is the chief source of common fractions. Recently a new folding decimal six-foot rule has been placed on the market. The new rule is graduated in the old units on the back and in feet, 0.1 feet, and 0.01 feet on the front. If this new rule is accepted by carpenters there will be a considerable decrease in the use of common fractions.

It is evident from the illustrations given above that a new era has arrived for decimals and with it new opportunities for the slide rule as an important and inexpensive instrument of computation.

# CHAPTER IV

## APPROXIMATE COMPUTATION AND THE SLIDE RULE

THE slide rule is an instrument for rapid approximate computation. If the students who are beginning the study of the slide rule are not acquainted with the principles and rules of approximate computation, the teacher should first give them the necessary instruction on this topic.

The treatment of approximate computation given below is confined to the work that is essential for those who wish to study the slide rule. Readers who wish a more complete treatment of approximate computation are referred to the books on this subject by Bakst,[1] Scarborough,[2] Tuttle and Satterly,[3] and others.

*Approximate numbers* are numbers that differ slightly from some true value for which they are used. When we write $0.333 = 1/3$, $0.667 = 2/3$, $1.414 = \sqrt{2}$, or $3.14 = \pi$, the numbers that are expressed by decimals are approximate numbers and are not exactly equal to the true or exact values for which they are used.

All measurements are represented by approximate numbers, and most of the numbers taken from compound interest tables, square root tables, trigonometric tables, and the like are approximate numbers. As many of the approximate numbers mentioned above are met in the elementary school, and in the early years of high school, it is necessary that all pupils understand approximate numbers and learn the techniques needed for computation with them.

If a distance has been measured so precisely that it is known to

[1] A. Bakst, *Approximate Computation*, The Twelfth Yearbook of the National Council of Teachers of Mathematics. Bureau of Publications, Teachers College, Columbia University, New York, 1937.

[2] J. B. Scarborough, *Numerical Mathematical Analysis*. The Johns Hopkins Press, 1930.

[3] L. Tuttle and J. Satterly, *The Theory of Measurements*. Longmans, Green and Co., 1925.

be nearer to 698.4 ft. than it is to either 698.3 ft. or 698.5 ft., then each of the first three figures of the number is correct and the fourth figure may be in error by not more than 0.05 ft. As each figure of the above number has a definite meaning or significance, we say the number has four significant figures. On the other hand, if after measuring the distance, only a rough approximation is needed and the distance is called 700 ft., then the 7, being less than half a unit in error, is significant, but the zeros are not. If the distance rounded to 700 ft. is actually between 699.5 ft. and 700.5 ft., then the zeros show the actual lack of quantity in the places in which they stand, and the number 700 ft. has three significant figures. The measurement 999.99 ft. rounded to four significant figures is 1000. ft. All the zeros in the rounded measurement are significant.

Most textbooks in mathematics state that to have all of its figures significant, *every figure of a number except the last must be correct and the error in the last figure must not be more than one-half unit in its place.* However, in science and practical work the usual practice is to retain figures which are in doubt by as much as 5 or 6. This often results in the retention of one more figure than would be retained according to the rule given above.[4]

The rules used in scientific work for determining the number of figures to be retained in the sum, difference, product, quotient, or root from the number of significant figures given in the data are in principle the same as those given later in this study.

When zeros are the result of correct measurement, they are significant. That is, if a distance is measured and found to be 2000.5 ft., the three zeros are significant. If a distance of 697 ft. were called 700 ft. as an approximation, then the first zero would be significant, but the second zero would not be significant. Similarly, the number 6000 ft. may have one, two, three, or four significant figures. Zeros are not significant when they are used simply to locate the decimal point in a rounded number, as in 7000 mi. Likewise, zeros used before some other digit in a decimal fraction to give place value are

[4] For details concerning the practice followed in the experimental sciences with reference to significant figures, see pp. 14 and 15 of *Precision of Measurements and Precision Problems* by G. V. Wendell and W. L. Severinghaus. Published by the Department of Physics, Columbia University, New York, 1938.

not significant; for example, the zeros in 0.00036 in. are not significant. These two uses of zero are likely to cause a certain amount of confusion, but usually too much emphasis is placed on them. There is one chance in ten that a two-figure measurement will end in zero; one chance in a hundred that a three-figure measurement will end in two zeros; and only one chance in a thousand that a four-digit measurement will end in three zeros. Consequently, there is little chance in practical work of meeting a measurement with three or four significant zeros.

All ambiguity with respect to the significance of zero may be removed by using standard numbers or the standard form of a number. In the standard number $6.00 \times 10^4$ the two zeros following the 6 are significant. In the standard number $8.63 \times 10^8$ the first three figures are significant, and in the standard number $8.6300 \times 10^8$ the first five figures including the two zeros are significant.[5] Lower case zeros may be used to indicate that the zeros are not significant. In the number 600oo. the first two zeros are significant. In the number 86300oooo, the first five figures including the first two zeros are significant.

In a set of numbers one can usually tell when the zeros are significant. If the person making the measurements at the left was familiar with approximate computation, it may be assumed that all the zeros are significant. Like people, zeros may be judged *by the company they keep*. If one sees such numbers as 6,000,000,000,000 miles, or 4,000 miles, or 188,000 miles, one may usually assume that these are rounded numbers, and that the zeros are not significant; and if the numbers are not to be used in computation it really does not matter greatly whether the zeros are significant or not.

430.0 ft.<br>
487.3 "<br>
900.0 "<br>
838.4 "

In the following cases, the zeros are *not significant* and are used merely to locate the decimal point or show place value: 24,000 mi.; 186,000 mi.; 8,000 mi.; 93,000,000 mi.; 4,000 mi.; 240,000 mi.; 0.00386 in.; 0.0183 in.; 0.00083 in. When zeros are between other significant figures, or when zeros follow other significant figures after a decimal point they are always significant. In the following cases the zeros are all significant: 2.025 lb.; .8000 in.; 4003 ft.;

[5] A full treatment of standard numbers is given on pages 46 to 49 of this book.

4.600 ft. 300.06 mi.; .318309; 301.00 ft.. In 0.009300 km. and 0.00304000 km. all the zeros except the first three zeros on the left are significant.

Considerable confusion exists regarding the number of significant figures in connection with angle measurement, because of the fact that angles are not usually measured in tenths or hundredths of a degree. If angles were so measured, as shown below, there would be no confusion.

| 1 sig. fig. | 2 sig. fig. | 3 sig. fig. |
|---|---|---|
| 6.° | 24.° | 124.° |
| 3.° | 6.3° | 63.5° |
| 0.4° | 0.42° | 4.23° |

To find the number of significant figures when an angle has been measured in degrees and minutes, change the measurement to minutes. For example, 46°52′ = 2812′, which has 4 significant figures. Transits costing from $200 to $400 usually may be read to the nearest minute.

To find the number of significant figures when an angle has been measured in degrees, minutes, and seconds, change the measurement to seconds. For example, 32°16′24″ = 116184″, which has 6 significant figures. Very expensive precision triangulation theodolites usually may be read to the nearest 5 seconds. It should be noted that the second should not be used in high school textbooks.

There may also be confusion when denominate numbers are used in linear and other measurements. How many significant figures are there in the measurement 43 ft. 4 3/8 in. if the measurement is correct to the nearest 1/8 in.? Since the above measurement is equal to 4163 eighths of an inch, it contains four significant figures.

How many significant figures are there in 24 lb. 11 oz. correct to the nearest ounce? Since 24 lb. 11 oz. = 395 oz., the measurement has three significant figures. If the measurement 6 ft. is correct to the nearest 1/8 in. it should be written 6 ft. 0/8 in. to show that the measurement has three significant figures (576 eighths). The measurement 6 ft. has only one significant figure. The measurement 46° has only two significant figures, but the measurement 46° 0′ has four significant figures (2760′).

The examples given above show that the old measures are much less satisfactory than decimal measures in all work involving approximate computation.

Physical constants or data taken from tables will often contain more significant figures than are needed in a given problem. Many people have learned the value of pi as 3.1416 or 3.14159, but for most uses 3.14 is sufficiently accurate. This is the value of pi rounded off to three significant figures.

Computation frequently produces an answer that contains more figures than are warranted by the original data. In such cases, the answer should be rounded off so that unnecessary figures will not be retained. When a slide rule is used, it is impossible to carry out computations to a degree of accuracy not warranted by the original three- and four-figure data.

The following rules may be used for rounding off figures:

1. "*If a number, given to a certain number of significant digits, is to be rounded off to a stated number of significant digits, the digits that are to be dropped should be replaced by zeros. In the case when the digits that are to be dropped are located to the right of the decimal point, the use of the zeros is not necessary.*" [6]

2. *If the left-hand digit of the digits that are to be dropped is 5 or more, then the right-hand digit of the digits that are to be retained should be increased by 1; otherwise the right-hand digit of the retained digits should be left unchanged.*

In a set of measurements like those shown in A, each measurement should be made to the same unit; in this case the *tenth* of a foot is considered to be the unit since each measurement is made to the nearest *tenth* of a foot. In set A, the zeros in 12.0 ft. and 10.0 ft. are significant. In the same way, in set B, each measurement is recorded correct to the nearest *hundredth*, the unit in this case being the *hundredth* of a foot. In set B, all the zeros in 40.00 are significant.

| (A) |
|---|
| 16.8 ft. |
| 7.5 " |
| 12.0 " |
| 3.7 " |
| 10.0 " |

If a number of measurements, some of which are slightly larger and others slightly smaller than their true values, are added as in C, there may be a small error in the total. However, in all practical work it appears better to *retain the full answer 29.76 ft.* than to use

| (B) |
|---|
| 36.20 ft. |
| 40.00 " |
| 18.64 " |
| 2.07 " |

[6] See Bakst, *op. cit.*, pp. 40-41, also p. 275.

the rule sometimes given to "round off the last figure of the answer." This rule would leave the answer 29.8 ft.

For practical work in subtraction it is better to retain the full answer as shown in D.

| (C) |
|---|
| 8.46 ft. |
| 5.12 " |
| 9.37 " |
| 6.81 " |
| 29.76 ft. |

The accuracy of a computed result or measurement has no relation to the number of decimal places in it. That the accuracy of measurement is not a function of the location of the decimal point may be seen from the following: 3.5 mm. = 0.0000035 km. It is seen that 0.0000035 km. is the two-figure measurement 3.5 mm. changed to kilometers and that both numbers have two significant figures. The position of the decimal point depends on the unit of measurement. *Many elementary textbooks instruct the pupil to obtain answers correct to one, two, or three decimal places.* Such instructions are doubtless responsible for the fallacy that accuracy of the answer is determined by the number of decimal places. The best criterion for accuracy in ordinary computation, including slide rule computation, is the number of significant figures. As 3.5 mm. and 0.0000035 km. each have two significant figures, they are of the same degree of accuracy. A number having three significant figures, or three-figure accuracy, is more accurate than a number having two significant figures or two-figure accuracy. A number with four significant figures, or four-figure accuracy, is more accurate than a number having three significant figures or three-figure accuracy. (In this study the expression *N-figure accuracy* will be considered equivalent to the expression *correct to N significant figures.*) It should be kept in mind, however, that if two numbers have the same number of significant figures, the one beginning with the larger digit will be the more accurate. For example, 999 is more accurate than 101; likewise 0.999 is more accurate than 101, since the decimal point does not affect accuracy. If 1 is added to 999, the result, 1000, will have four significant figures, whereas if 2 is subtracted from 101, the result, 99, will have only two significant figures. Another way to show that 999 is more accurate than 101 is to compare their relative errors, as explained in the second paragraph below.

| (D) |
|---|
| 87.64 |
| 28.37 |
| 59.27 |

The fact stated in the last paragraph is of interest in connection with the slide rule. *At the left-hand end of a 10-inch rule, it is quite as easy to read the rule to four significant figures as it is to read the right-hand end to three significant figures.* This leads many people to believe that the right-hand end of the rule is less accurate than the left-hand end. However, there is very little difference in accuracy between the two ends of the slide rule since such numbers as 976 and 1056 differ but little in accuracy.

The best standard for accuracy is the *relative error* or *percentage of error*. The relative error is the ratio of the error in a measurement to the measurement itself. If this is expressed as a per cent, it is called the percentage error. If there is an error of 1 in the last figure of each of the numbers 976 and 1056, the relative errors will be 1 : 976 and 1 : 1056; the percentage errors will be 0.102% and 0.0947%. The number 1056 has a slightly smaller percentage error than 976 and so it is more accurate.

In finding the relative or percentage errors, it is usually sufficient to round the given number to two significant figures and then to divide the error by this rounded number. In so doing, the results can generally be obtained mentally, or a slide rule may be used.

The relative error in any number is often much more important than the absolute error, and in case of a change in magnitude the relative change is often much more significant than the actual change. A loss of twenty pounds in the weight of a two-ton elephant is of no significance, but the same change in a man weighing 120 pounds would be serious, while in the case of a 40-pound child it would doubtless prove fatal. The actual change is the same in each case, but the relative changes are 20 : 4000, 20 : 120, and 20 : 40 or 1/200, 1/6, and 1/2.

It should be noted that in addition to the approximate computation needed in connection with the slide rule, the pupil meets approximate computation in connection with most pencil solutions. Graphic solutions are also approximate. Usually only two or three significant figures are obtained in graphic solutions. The use of square root tables, trigonometric functions, and logarithms introduces other approximate three-, four-, and five-figure solutions. A study of approximate computation should precede all the above topics.

When two approximate numbers are to be multiplied, the following rules should be used:

1. *If two approximate numbers have the same number of significant figures, multiply the numbers and round off the product to the same number of significant figures as there are in each factor.* The last figure of the answer will not always be significant, but this rule is satisfactory for all elementary work and is usually followed in scientific work.

2. *If two approximate numbers are such that one of them has more significant figures than the other, first round off the more accurate approximate number so that it contains one more significant figure than the less accurate approximate number. Then multiply the numbers and round off the product to the same number of significant figures as there are in the less accurate factor.*[7] A product cannot be more accurate than the least accurate of the factors in the computation.

The following examples illustrate these rules:

Example (1): What is the area of a rectangle 8.46 ft. by 9.62 ft.?

Solution: 8.46 × 9.62 = 81.3852 sq. ft. This answer should be rounded to 81.4 sq. ft.

Example (2): What is the area of a rectangle 684.384 ft. by 6.86 ft.?

Solution: 684.4 × 6.86 = 4694.984 sq. ft. This answer should be rounded to 4690. sq. ft. The zero is not significant.

When two approximate numbers are to be divided, the following rules should be used:

1. *If the two approximate numbers have the same number of significant figures, divide the numbers and carry the quotient to one more significant figure than is contained in each of the given numbers. Then round off the quotient so that it contains the same number of significant figures as the given numbers.*

2. *If the dividend and divisor are such that one of them has more significant figures than the other, first round off the more accurate number so that it contains one more significant figure than the less accurate number. Then divide the numbers and carry the quotient to one more significant figure than is contained in the less accurate of the given num-*

[7] Rules similar to those given above may be found in Bakst, *op. cit.*

*bers. Finally round off the quotient to the same number of significant figures as are found in the less accurate of the given numbers.*

The examples below illustrate the two rules for division of approximate numbers given above. The similarity of the rules for multiplication and division should be noted.

Example (1): What is the quotient of 9.76 ÷ 3.24?
Solution: 9.76 ÷ 3.24 = 3.012. This answer should be rounded off to 3.01.)

Example (2): What is the quotient of 864.39 ÷ 3.87?
Solution: 864.4 ÷ 3.87 = 223.3. (Should be rounded off to 223.)

When the above rules for dividing approximate numbers are used, the last figure of the quotient will not always be significant. In the actual work of dividing two approximate numbers, it is often necessary to annex zeros to the dividend in order to secure the number of figures in the quotient warranted by the original data. It is never permissible to annex zeros in addition and subtraction since this changes the accuracy of the original data. In division it is the result rather than the original data that is affected.

When used in multiplication and division, the rules of approximate computation should be applied only to approximate factors. If a formula contains exact numbers as well as approximate numbers, the rules of approximate computation do not affect the exact numbers. When a formula contains approximate numbers such as pi or the tangent of an angle, the values used for pi or the tangent should be rounded off to one more significant figure than is contained in the least accurate of the other approximate numbers in the formula. This harmonizes with the rules given above. Examples illustrating these procedures are given below:

Example (1): What is the volume of a cone 32.76 ft. in diameter and 49.41 ft. high?
Formula: $v = 1/3\ \pi\ r^2\ h$
Solution: $v = 1/3 \times 3.1416 \times (16.38)^2 \times 49.41 = 13882.4.$

This answer should be rounded to 13880. The 3 in the above formula is an exact number. Since there are four significant figures in the measurements, pi is rounded to five significant figures.

Example (2): Find the diameter of a circle whose circumference is 47.36 ft.

Solution: Diameter = 47.36 ÷ 3.1416 = 15.075. The quotient should be rounded to 15.08.

The value of pi is selected as 3.1416 since it should have one more significant figure than 47.36.

Example (3): Divide 49.68 by 8.3.

Solution: Following Rule 2 for division this becomes 49.7 ÷ 8.3. Since two zeros have to be annexed to 49.7, thus making it 49.700, it seems equally sensible in such a case to use 49.680 instead of 49.700 as the dividend. The quotient carried to three significant figures is 5.98 in either case. This result is finally rounded to 6.0.

In square root, as in division, the answer should contain in general the same number of significant figures as the measured data. Here again, as in division, it may be necessary to annex zeros. In extracting the square root of 86.3 on paper it is necessary to annex three zeros to secure three figures in the answer. When the slide rule is used, it is possible to secure a three-figure answer without annexing zeros. The square roots of 64 sq. ft., 6400 sq. ft., 81 sq. ft., and 8100 sq. ft. (all the figures including the zeros are significant) are usually given as 8 ft., 80 ft., 9 ft., and 90 ft. When used in practical problems, these square roots should be 8.0 ft., 80.00 ft., 9.0 ft., and 90.00 ft.; otherwise there would be a loss in the number of significant figures. This loss in significant figures might be serious, especially where computation follows the extraction of square root. *It is better to carry results in square root to one more figure than the number warranted by the data and then round back.*

Approximate computation is necessary in many phases of high school mathematics, such as square root, radicals, quadratic equations, numerical trigonometry, all graphic work, all practical problems based on measuring, and with logarithms. It is also necessary in almost all work in science. In spite of this fact there are very few high school textbooks that give an adequate treatment of this very important topic. Even those that do give some attention to it often negate much of what they have done by the model solutions and answers that they give. Duane Roller says: "Most textbooks discuss the concept of significant figures but fail to take it into account in

stating problems of a quantitative nature, thus making it almost impossible for the student to use the idea of significant figures in solving problems." [8]

A number of our better high school textbooks have many problems containing data given correct to one, two, or three significant figures. However, the answers to these problems, as given in the answer books or keys, may have five, six, or even more supposedly significant figures. For example, one high school textbook has a problem in which a ship sails 4.0 miles. This fact and the size of the two including angles measured to two significant figures constitute the data. Another side is wanted. The answer given is 8.4746 miles Since the data in the problem are correct to only two significant figures, the only sensible answer to this problem is 8.5 miles.

*No textbook should give, and no teacher should allow, more significant figures in a final answer in multiplication, in division, or in square root than there are in the least accurate item used, and all texts should show by the data used the accuracy desired in the answer. All data in a given problem should be consistent.*

The computational habits of pupils and even of teachers are fixed by the answers given in answer books, specimen solutions, and teachers' keys. A pupil works hundreds of problems and naturally tries to get the answers given in the answer book. A text may give fairly satisfactory directions for approximate computation, but if it fails to back these up by its answers, the directions will have little effect.

If the suggestions given in this chapter are followed, it will be much easier to teach pupils to use the slide rule, and there will be no difficulty in determining, from the data given, which problems may be solved on the slide rule. It is an interesting fact that pupils who use the slide rule almost automatically make use of the proper rules for approximate computation with three- and four-figure data.

[8] Duane Roller, *Terminology of Physical Science*, p. 36. University of Oklahoma Press, 1936.

# CHAPTER V

## ACCURACY AND THE SLIDE RULE

In teaching the use of a slide rule a serious psychological difficulty is met. In all their previous training, the students have been accustomed to carry results to six, eight, or more supposedly significant figures. Because of this conditioning, they feel that the results obtained by the use of the slide rule are not sufficiently accurate to be of much value. A similar difficulty applies to graphic solutions, scale drawing, and to some extent to tables, such as those for square and cube root. In practical work in mathematics and science, students are often discouraged when they find they can get only three-figure or four-figure accuracy. These same students, however, are quite willing to submit an answer containing six significant figures to the following problem:

> A tank contains 279 cu. ft. of water. Find the weight of this water in pounds if 1 cu. ft. water = 62.5 lb.

The answer that these students give to this problem is 17437.5 lb. but it should be 17400 lb., a number which has only three significant figures.

The slide rule is an instrument for rapid approximate computation. If a problem is solved on a slide rule, an answer having two, three, or four significant figures may be obtained. The number of significant figures will depend on the type of slide rule used and the skill of the operator.

*It is useless to attempt to teach the slide rule until thorough respect for a three-figure answer has been developed.* Few students appreciate the relative importance of the various digits of a number. It is unfortunate that instruction on this topic is not given in the elementary school; if it were, much time would be saved later on in high school in correcting false impressions concerning this matter. Approximate computation may be taught in the seventh grade.

An excellent way to get a vivid impression of the relative importance of the digits of a number is to draw on the blackboard a graph similar to the one which is shown in Fig. 1. This graph illustrates the value of each of the digits of the number 8888. If a line 10 ft. long represents the first 8 on the left, a line 1 ft. long will represent the second 8, a line 0.1 ft. long the third 8, and a line 0.01 ft. long the fourth 8. It is evident from Fig. 1 that the bar representing the fourth 8 is very small and unimportant in comparison with the first bar. The graph also shows that the third 8 has little importance in comparison with the first and second 8's. The graph emphatically shows that if the first three digits are correct, the fourth and all subsequent figures will have very little relative importance. In slide rule work, errors of 1 unit in the third place are common. If 8870 or 8890 were graphed in Fig. 1 in place of 8880, the change in the length of the third bar would be too small to be noticed.

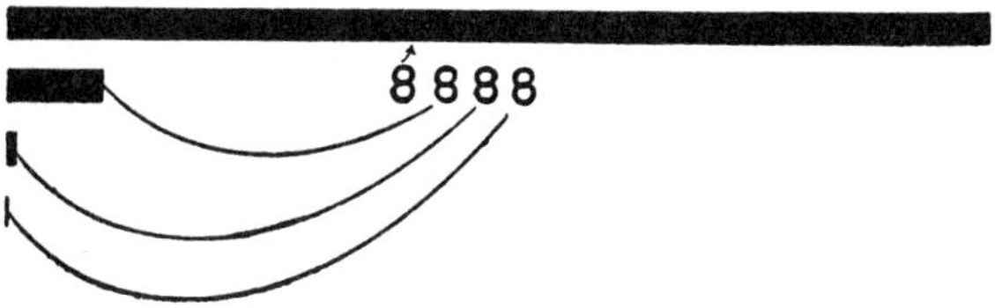


Figure 1

Figure 2

If Fig. 1 had attempted to represent a six-digit number and the sixth (last) digit were represented by a bar 0.01 ft. long, then the other five digits would be represented by bars of the following lengths: 0.1 ft. (fifth digit); 1 ft. (fourth digit); 10 ft. (third digit);

100 ft. (second digit); 1000 ft. (first digit). It is apparent that the bars representing the first and second digits would be much too long to be drawn on the blackboard.

The graph in Fig. 2 compares the first three digits of the number 666666 with the last three digits. On the blackboard the line representing 666000 would be 11.1 ft. long if the line representing 666 were 0.0111 ft. long (or approximately 0.01 ft.). Fig. 2 is more striking, perhaps, than Fig. 1 for the purpose of emphasizing the relative unimportance of the last digits of a large number.

A discussion such as that given above will help to convince students that for many practical purposes three- or four-figure accuracy is sufficient, and that the relative value of all digits beyond the fifth place is very slight indeed.

In an eighth grade arithmetic class visited recently, the teacher had each of the 32 pupils measure five different circular objects, divide the circumference by the diameter in each case, and find the average of the five results. She then found the average of the 32 averages. The result rounded to three figures was 3.14. Some of the pupils thought that she should have gotten 3.1416, but the 3.14 was as satisfactory as one should expect. Experiments of this kind show how hard it is to get more than three figures correct by measuring.

An experiment in finding the values of the natural tangents of angles by measuring the sides of right triangles and dividing their lengths will show that it is very difficult to get the third digit correct.

The accuracy of a 10-inch slide rule is about 1 in 1000. On a 20-inch rule it is possible to obtain answers correct to about 1 in 2000; this fully insures the accuracy of the third digit and allows a skillful operator to approximate the fourth place. The Fuller, Thatcher, and other large slide rules give results correct to within 1 in 10,000, or full four-figure accuracy. Three- and four-figure results are sufficiently accurate for a large proportion of the computational needs of most persons. Where greater accuracy is desired, logarithms or computing machines should be used. The slide rule is not a substitute for either of these, but in its own field it is superior to them. Logarithmic solutions are slow and laborious in comparison with slide rule solutions, and the computing machine has the handicap of its size, weight, and high price.

In order to determine the degree of accuracy acceptable in industry, an analysis was made of the tolerances allowed in commercial weighing and measuring.

Table I on page 30 shows the tolerances approved by the Federal Government and practically all state governments. The word *tolerance* refers to the allowable amount of variation from a standard measure. The word *deficiency* as used in the table means over-registration of the measuring device; in such cases the customer is given "short measure" or "short weight." The word *excess* means under-registration, in which case the customer gets "over weight" and the seller loses.

The tolerance depends on the type of measuring instrument and the amount measured. Table I shows that for wood rules 1 foot to 6 feet long the tolerance is 1 in 384, but on 100-foot steel tapes it is only 1 in 4800. A tolerance of 1 in 4800 on a 100-foot steel tape means that a steel tape 100 ft. long must not have an error of more than 1 part in 4800 parts; hence in measuring a length 100 ft. long, the tape must give a result that is correct within approximately 0.02 ft.

In using the instruments cited in Table I a larger error than the tolerance is sometimes made through the user's carelessness. Errors greater than the tolerance may also be caused by the varying physical conditions of the thing measured. For example, if coal is wet before it is weighed, the weight will be increased from 5% to 20%, depending on the size and the quality of the coal. In using a steel tape it is possible to have errors from at least eight different sources, each error being as large as the tolerance. Because of the differences in the quality and the physical conditions of the substances weighed or measured it would be foolish to have closer tolerances than those given in the table.

Some of the difficulties in measuring wheat have been pointed out by Strayer and Upton in one of their textbooks.[1] Only a few of the possible sources of errors, however, are given in their discussion.

[1] G. D. Strayer and C. B. Upton, *Practical Junior Mathematics, Book Two*, p. 73, American Book Co., New York.

Table I

Official Tolerances for Measuring Instruments

| Measuring Instrument | Size or Capacity of Instrument | Tolerance Deficiency—One in | Tolerance Excess—One in |
|---|---|---|---|
| Wood rules | 1 to 6 ft. | 384 | 384 |
| Steel tapes | 3 ft. | 1152 | 1152 |
| | 10 ft. | 1920 | 1920 |
| | 100 ft. | 4800 | 4800 |
| Fabric measuring devices | 1 yd. | 144 | 96 |
| | 5 yd. | 480 | 288 |
| | 7 yd. | 504 | 252 |
| Capacity measures: tin, copper, etc. | 1 qt. | 115 | 64 |
| | 1 gal. | 257 | 128 |
| | 5 gals. | 214 | 105 |
| | 10 gals | 257 | 128 |
| Milk bottles | 1 pt. | 96 | 42 |
| | 1 qt. | 116 | 64 |
| Grease measuring devices | 1 gal. | 77 | 77 |
| Vehicle tanks | 50 gals. | 100 | 100 |
| | 175 gals. | 175 | 175 |
| | 1325 gals. | 221 | 221 |
| Dry measures | 1 pt. | 34 | 17 |
| | 1 qt. | 46 | 23 |
| | 1 peck | 68 | 34 |
| | 1 bu. | 86 | 43 |
| Counter scales | 1 lb. | 256 | 256 |
| | 5 lbs. | 426 | 426 |
| | 20 lbs. | 640 | 640 |
| | 240 lbs. | 960 | 960 |
| Spring scales | 1 lb. | 128 | 128 |
| | 12 lbs. | 192 | 192 |
| Wagon and truck scales | 500 lbs. | 400 | 400 |
| | 1500 lbs. | 500 | 500 |
| | 10000 lbs. | 500 | 500 |
| Gold coins | $ 5 | 516 | |
| | $10 | 1032 | |

The finest grade of clean wheat should weigh about 50 lbs. per cubic foot; good commercial wheat will weigh 48 lbs., while wheat of poorer quality may weigh as little as 40 lbs. Old wheat will not weigh as much as new wheat. The variations in the quality and age of wheat may easily introduce errors in measuring that are much greater than the official tolerances of the measuring instruments.

Strayer and Upton in the textbook cited above came to the conclusion that the error in measuring grain will seldom be less than 1 in 200. This justifies the use of the approximate equivalent of 1 bu. = 1.25 cu. ft., or 1 cu. ft. = 0.80 bu., in place of the cumbersome but more exact value, 1 bu. = 2150.42 cu. in. These approximate equivalents are in error by less than 1 in 200. However, the tolerance allowed by law is from 1 in 43 over, to 1 in 86 under, on bushel measures; so a tolerance of 1 in 200 is certainly sufficiently conservative. Computations performed on a 10-inch slide rule will have errors of only 1 in 1000.

Tolerances are also recognized in everyday business computations though the word *tolerance* is not officially used in connection with such practices. When a businessman, in computing interest, counts the year as 360 days instead of 365 days, he is allowing a tolerance of 1 in 73.

It would appear that the slide rule could be used more extensively in business calculations than is now done. If a tolerance of 1 in 1500, which is far less liberal than that allowed in commercial measuring, were universally adopted for business computations, the 20-inch slide rule could be used for such calculations. This would not increase the errors in business calculations since no answers can be more accurate than the data used; most of the data obtained from weighing or measuring will have far larger errors than those that will be introduced by the slide rule.

G. E. Lundell[2] of the Bureau of Standards, Washington, D. C., writes: "As for estimating the accuracy of his work, it can be said the determinator is usually an optimist in thought and expression, while the analyst is a confirmed pessimist. The determinator reports silica in glass as 71.61, if not 71.611. The analyst, who knows

[2] G. E. Lundell, "The Chemical Analysis of Things as They Are," *Industrial and Engineering Chemistry* (Analytical Edition), Vol. 5, No. 4, July 15, 1933.

that he is doing nicely to insure results in the first decimal place, reports 71.6 and thus is honest with himself and deceives no one as to his powers."

One of the tables given by Mr. Lundell is interesting in shedding light on the accuracy to be expected in the commercial laboratory. This table (part of which is reproduced here as Table II) shows the

TABLE II

Average Percentage Errors of Determinations of Constituency of Representative Materials (abridged) *

| Constituent | Material | Per Cent of Constituent | Average Per Cent of Error |
|---|---|---|---|
| Si | Plain steel | 0.39 | 0.86 |
| Si | Tungsten steel | 0.48 | 2.6 |
| Si | Ferrosilicon | 75.6 | 0.2 |
| $Al_2O_3$ | Glass | 0.33 | 14.2 |
| $Al_2O_3$ | Limestone | 4.16 | 1.3 |
| $Fe_2O_3$ | Soda-lime glass | 0.06 | 9.5 |
| Fe | Manganese bronze | 1.13 | 1.7 |
| $Na_2O$ | Plastic clay | 0.27 | 18.8 |
| $K_2O$ | Silica brick | 0.29 | 4.7 |
| Sb | Cast bronze | 0.16 | 10.4 |
| Cu | Cast iron | 0.151 | 5.4 |

* Adapted from Lundell, *op. cit.*

average percentages of errors in determining the constituency of various materials. In connection with this table Lundell says: "The ordinary analyst of things as they are is not so much concerned with keeping his errors below 1 part in 1000 as with keeping them below 1 in 100." Also, "An error of 0.01% in determinations of carbon in a 0.20% carbon steel, represents an error of 5% and is yet within the accuracy of ordinary sampling and any use that can be made of the analysis in the ordinary case.

"Such errors are not confined to chemical analysis alone. They apply equally well to estimations made by physical methods, such as the spectroscope."

For all the computations reported in Table II a 20-inch slide rule would have been satisfactory, and, for some of it, a 5-inch rule would have been sufficiently accurate.

In writing on the topic, "What Is Scientific Exactitude?" T. Swann Harding[3] says: "If in reading a polariscope (or any instrument of precision), or in making my final weighing incidental to a phosphorous determination, I managed to get four or five results which were exactly alike, I at once became suspicious of myself. I felt certain that something was wrong. Instead of such exactitude, I expected to find errors of 5% not at all unusual. To secure what science calls *exactitude*, I averaged these results and took the final average as correct."

From the above quotations, which have been selected from a large number of similar ones as fairly representative, there is but one conclusion. In all this work the practical analyst would be happy indeed if he could always be sure that the first three figures of his results were absolutely correct. A 20-inch slide rule would be sufficiently accurate for any computation with such results.

The following proportion is taken from the *Chemist Analyst* of September, 1932:

$$55.84 : 16.03 = 0.005 : x$$

The answer given is 0.00144. The answer obtained on a 20-inch slide rule is 0.001435. The answer obtained by actual pencil computation is 0.0014352; this answer and the slide rule answer would both be rounded to 0.00144.

An analysis of a large number of similar equations showed that the 20-inch slide rule was sufficiently accurate for 92% of the equations and for most of the remaining 8%.

As an additional check on the number of significant figures used in fairly technical work, the author assembled 3479 problems from three different sources[4] and classified them according to the number

[3] T. Swann Harding, "What Is Scientific Exactitude?" *Atlantic Monthly*, April, 1931, pp. 514-523.

[4] The Analytical Edition of *Industrial and Engineering Chemistry*, Vol. 7, 1935.

L. C. Urquhart and C. E. O'Rourke, *Design of Concrete Structures*. McGraw-Hill, New York, 1935, 556 pp.

Charles D. Conklin, Jr., *Structural Steel Drafting and Elementary Design*. John Wiley & Sons, Inc., New York, 1915, 154 pp.

of significant figures in the *largest number* used in each problem. (The phrase *largest number* means the number with the greatest number of significant figures.) The 3479 problems were classified as in Table III.

TABLE III

Number of Significant Figures in the Largest Factor in Each of 3479 Technical Problems

| Number of Significant Figures | Number of Problems | Per Cent | Cumulative Per Cent |
|---|---|---|---|
| 2 | 805 | 23.1 | 23.1 |
| 3 | 1685 | 48.5 | 71.6 |
| 4 | 818 | 23.5 | 95.1 |
| 5 | 156 | 4.5 | 99.6 |
| 6 | 15 | 0.4 | 100.0 |
| Total | 3479 | | |

A number of problems classified as having six significant figures were of the following type, $m = \frac{2.23770 - 2.21215}{2.650 - 2.150}$. There were seven of these problems on a single page. While there are six significant figures in each of the n ımbers in the numerator, the expression actually equals $\frac{0.02555}{0.500}$, and the answer is 0.0511. This answer could have been obtained on either a 10-inch or a 20-inch slide rule.

More than 70% of these 3479 problems could have been solved on a 10-inch slide rule, more than 95% on a 20-inch rule, and 100% on a Thatcher or Fuller slide rule.

If, in one of the above problems, a measurement with three significant figures was used with one having five significant figures, the problem was classified with those having five significant figures because it was not always possible to tell whether a number was a result of measurement or an exact number. In view of this, the claims made above for the slide rule are conservative.

D. S. Davis[5] says, "The convenient 20-inch desk type slide rule is nearly as good as a table of four-place logarithms and can be used for most practical work."

Thousands of physical and mathematical constants are used in scientific and industrial work. Many of these constants have been determined to four, five, or even six significant figures but there are others that have been determined to only three or four significant figures. For example, until recently the constant of gravitation was known with an accuracy of only 1 part in 600. Recent experiments carry the precision to about 1 part in 6000. Results as accurate as this may be obtained on a Thatcher slide rule or on other precision slide rules.

It is interesting to know the number of significant figures used in very accurate scientific or commercial work. One of the most refined series of measurements ever attempted was the measuring of the length of the meter in terms of wave lengths of red cadmium light by Michelson (1892–1893) and Benoit, Fabry, and Perot (1906). The results were subjected to corrections for humidity, temperature, and pressure, and were the means of three and seven experiments respectively. They were as follows:

| | |
|---|---|
| Michelson | 1553164.03 |
| Benoit, Fabry, and Perot | 1553164.13 |

Here is perfect agreement to seven significant figures.

The base line used by Michelson[6] in his experiment with the velocity of light (1925) was considered correct to six significant figures, 35385.50 m. This base line is considered the most accurately measured geographic distance in the world. The velocity of light in vacuo was calculated to be 299796 km. per second (186284 mi. per sec.); Michelson considered this velocity to be correct to five and possibly to six significant figures.

It may be possible, with the utmost care, the finest instruments, and the help of statistics, to measure to six, seven, or even eight sig-

[5] D. S. Davis, "Precision of Measurements," *The Chemist Analyst*, Vol. 25, No. 4, October, 1936, p. 78.

[6] A. A. Michelson, "Measurement of the Velocity of Light," *Contributions from the Mount Wilson Observatory*, No. 329, 1927.

nificant figures. Such measures are, however, the extreme limit of scientific accuracy. Problems using such highly accurate data are *rare*, but may be ***extremely important.*** There are also many places in commercial calculation where highly accurate data are used. Commercial tables often give 10- or 12-figure data, as $(1.08)^{100} = 2199.76125634$.

The slide rule cannot be used with data having six, seven, or eight significant figures, but the expert using such data may find many uses for the slide rule.

The value of the slide rule as an instrument for checking solutions in trigonometry, chemistry, physics, and other subjects is fully recognized. Certainly an instrument sufficiently accurate for 60% to 90% of the original solutions in *commercial physics* and *chemistry* is sufficiently accurate for checking such solutions.

The slide rule is of maximum value with formulas in which there is no addition or subtraction, such as the law of sines $\left(\frac{a}{\sin A} = \frac{b}{\sin B} = \frac{c}{\sin C}\right)$. It is of limited value with formulas that combine addition and subtraction with other operations, like the law of cosines, $\cos A = \frac{b^2 + c^2 - a^2}{2\,bc}$.

The above limitations of the slide rule should be kept in mind. Excessive claims for the slide rule will do more harm than good.

# CHAPTER VI

## COMMON ERRORS MADE IN SLIDE RULE COMPUTATION

This chapter presents the results of a study of the errors that were made by students in learning the use of the slide rule. A knowledge of the type and relative frequency of these errors is prerequisite to the development of proper teaching procedures.

The author gave a simple test in multiplication (Test A below) to 660 students who had previously received instruction in the use of the slide rule during six regular class periods of 50 minutes each. These students were asked to spend two hours practicing slide rule computations for each period of class instruction; hence they had had a total of 17 hours of instruction and practice before taking the test.

The 660 students taking the test included 554 students from various mathematics classes in Teachers College, Columbia University, 73 students from mathematics classes at State Teachers College, Trenton, N. J., and 33 students from mathematics classes in Hunter College, New York City.

Test A consists of simple examples which were so selected that it would be possible to determine the type of error for each solution. These examples were as follows:

### TEST A

Work the following examples rapidly but carefully. Do all the work on the slide rule. Do not check on paper. Write your answer after each example.

| | |
|---|---|
| 1. 128 × 161 | 6. 0.00463 × 0.0366 |
| 2. 1.65 × 1.95 | 7. 8.15 × 1.24 |
| 3. 135 × 670 | 8. 212 × 2.64 |
| 4. 96.5 × 3.94 | 9. 1.005 × 1.84 |
| 5. 80.5 × 6.15 | 10. 8.15 × 2.46 |

TEST A—*Continued*

11. 3.27 × 6.35
12. 4.65 × 8.82
13. 44600 × 5250
14. 0.00778 × 46500
15. 0.00473 × 0.00575

Some typical errors made by students working Test A are illustrated below by various incorrect answers to example 1. The correct answer to example 1 is 20600.

| Incorrect Answers | Type of Error |
|---|---|
| 1. 2060 and 206000 | 1. Locating the decimal point |
| 2. 26000 | 2. Reading 260 for 206—a zero error |
| 3. 2600 and 260000 | 3. Combination of 1 and 2 |
| 4. 20300 | 4. Assigning wrong value to tertiary graduations |
| 5. 2030 and 203000 | 5. Combination of 1 and 4 |

Various other errors in example 1 were caused by incorrect location of the given factors on the scales.

A total of 2853 errors were made by the 660 students in working the 15 examples in Test A. The tabulation and classification of these errors is presented in Table IV.

*Errors in Locating the Decimal Point* (Item A, Table IV)

There were 952 decimal errors, which were 33.4% of all the errors made on Test A. These figures clearly demonstrate the need of using a reliable method for locating the decimal point. Without such a method the slide rule is of little value. The 660 students had been taught to locate the decimal point in the answer by inspection or by a rough mental calculation. For example, in finding the product of 1.28 × 1.61 the student reads 206 from the slide rule and easily sees by inspection that the decimal point follows 2, giving 2.06 as the result. In finding the product of 128 × 161, the operator roughly approximates the product by thinking, "100 × 200 = 20000," which tells him that the product has five figures. The slide rule reading which was 206 is then changed to 20600. In examples like this it is not always easy for the operator to tell quickly the number of figures in the product unless he has a dependable method of locating the decimal point. The inspection method which has been

TABLE IV

Types of Errors Made by 660 Students on Test A

| Type of Error | Number of Errors | Per Cent of All Errors |
|---|---|---|
| A. Error in placing decimal point. "Decimal error." | 952 | 33.4 |
| B. Error in estimating third significant figure. (Ex. 856 for 858) | 723 | 25.4 |
| C. Omitting zero in second place. "Zero error." (Ex. 26 for 206) | 599 | 20.9 |
| D. Error in reading the tertiary graduation. (Ex. 52.1 for 52.5; 321 for 322) | 389 | 13.6 |
| E. Reading tertiary graduation for secondary. (Ex. 240 for 234) | 52 | 1.8 |
| F. Error in counting graduations. (Ex. 378 for 388) | 51 | 1.8 |
| G. Omitting initial 1 between primary 1 and 2. (Ex. 36 for 136) | 37 | 1.3 |
| H. Taking both factors on same scale* | 15 | 0.5 |
| I. Missing first significant figure (Ex. 443 for 543) | 8 | 0.3 |
| J. Repeating figures (Ex. 332 for 32) | 7 | 0.3 |
| K. Mistaking 3 for 8, 3 for 5, or 5 for 8 | 6 | 0.2 |
| L. Using one factor twice | 6 | 0.2 |
| M. Putting 1 before some number after primary 1. (Ex. 1246 for 246) | 5 | 0.2 |
| N. Dropping first significant figure other than 1. (Ex. 340 for 234) | 3 | 0.1 |
| Total | 2853 | 100.0 |

* Placing the index over the first factor and then reading this factor for the answer.

illustrated is of rather doubtful value in solving complex examples.

Many mistakes were made by the students in the relatively simple examples on Test A. The large number of decimal point errors in Test A led the author to adopt the *standard number method* of locating the decimal point. This method is described in Chapter VII.

*Errors in Estimating the Third Significant Figure* (Item B, Table IV)

The mistakes in estimating the third significant figure are not as serious as the errors listed under A, C, and D above. Most of the mistakes are made with numbers which are between the primary graduations 4 and 10. The greater number of errors made in this part of the scale is due to the smaller number of tertiary graduations and the consequent necessity for estimating the third significant figure. An error of one unit in the third significant figure of a number between the primary graduations 4 and 10 is equivalent to an error of from 1/4 of 1% to 1/10 of 1%. An error of two units in the third figure of the answer is equivalent to an error of 1/2 of 1% to 1/5 of 1%.

With practice it is possible to get the third significant figure correct about 7 times out of 10. This result was obtained by the author in the process of experimentation in the preliminary stages of this study. In this initial experimentation 2000 examples in reading square roots on the slide rule were solved by a group of high school pupils who were known to have superior ability in working with the slide rule. The slide rule answers obtained by these pupils were checked against a table of square roots, the values given in the table being rounded to three significant figures. In this way it was found that the slide rule answers had the third significant figure correct in 7 cases out of every 10.

A somewhat similar experiment with multiplication with two factors was performed by Runge and Koenig.[1] The products were read on the C and D scales of the slide rule and were checked against machine multiplication. The average percentage of error of the slide rule products obtained by these expert operators was 0.06%. This is equivalent to an error of about 0.15 mm. on a 25 cm. slide rule (10-inch rule). This in turn signifies that readings on the slide rule generally give results correct to three or four significant figures.

The methods used by the author for reducing the number of errors made in connection with estimating the third significant figure are given in Chapter VII, pages 62–64 of this study.

*Zero Errors* (Item C of Table IV)

In setting or reading numbers on the slide rule, students often omit a zero when it is the second significant figure of a number and

[1] C. Runge and H. Koenig, *Vorlesungen über Numerisches Rechnen*, pp. 10-12, 85-86, 330-331. Julius Springer, Berlin, 1924.

write the third significant figure in its place. For example, 175 may be set or read for 1075, or 260 may be set or read for 206. Similar mistakes may be made when the third significant figure of a number is zero. Zero errors in Test A account for about 21% of the total number of errors. Zero errors in the second place are very serious because they result in a large percentage error; if 180 is taken for 108, the error is over 65%. Exercises designed to reduce the number of zero errors are given in Chapter VII, pages 64 and 65.

*Errors in Reading Tertiary Graduations* (Item D, Table IV)

Students frequently make mistakes in reading the tertiary graduations on the C and D scales; that is, a student may read 203 for 206, 321 for 322, 394 for 398, or 431 for 435.

Errors in reading tertiary graduations may cause errors of several units in the third significant figure. Such errors are less serious than zero or decimal point errors, and are more serious than errors in estimating the last digit. Errors in estimating the last digit usually are not larger than one unit in the third significant figure. Errors in reading tertiary graduations may be reduced by a special technique given in Chapter VII, pages 60–62.

*Omitting the Initial 1 between the Primary Graduation 1 and 2* (Item G, Table IV)

There were only 37 of these errors. This small number is probably due to the fact that it was generally easy to determine the first significant figure of the answer by inspection. More frequent omissions of an initial 1 should be expected in complex examples. On a 20-inch slide rule more of the secondary graduations are numbered than on a 10-inch slide rule; hence students are more likely to omit the first figure of the answer when they use the 20-inch slide rule. A student who reads 9650 instead of 1965 makes a very serious error. Adequate drill in reading and locating numbers between the primary graduations 1 and 2 will, to a considerable extent, prevent the omission of the initial 1. There were only three errors of the closely related type N, dropping the first significant figure when it was 2 or more. There were five errors of type M, putting 1 before some number.

### *Miscellaneous Errors* (Items E, F, H, I, J, K, L, M, and N, Table IV)

The miscellaneous errors classified in items E, F, and H to N of Table IV are probably due in great part to speed, poor eyesight, and carelessness. Many students had slide rules with the old type of framed indicator, which covered up part of the scale and increased the number of errors. Fortunately these miscellaneous errors occurred in relatively small numbers.

### *Methods of Preventing Errors*

After giving Test A the author experimented for two years with various methods of teaching slide rule computation and finally worked out methods that seemed to reduce greatly errors of types A, B, C, and D. These methods are described in Chapter VII.

# CHAPTER VII

## METHODS OF PREVENTING SLIDE RULE ERRORS

### PREVENTING ERRORS IN LOCATING THE DECIMAL POINT

MANY teachers of mathematics give liberal credit to answers with the correct significant figures but with the decimal point incorrectly located. However, students should be taught that the most serious mistake they can make is that of misplacing the decimal point, and no credit should be given for any answer that has the decimal point in the wrong location.

If the correct answer to a problem is 602 ft., and a student gets 598 ft. for his answer, he has every significant figure of the answer wrong, yet his answer has an error of less than 1%. An error of this size would cause very little trouble in most practical work. On the other hand, any mistake in the location of the decimal point introduces an enormous error. Thus, if 60.2 ft. is given as the answer instead of the 602 ft., there is an error of 90%; if 6020 ft. is given, the error is 900%; and such answers as 6.02 ft., 0.602 ft., or 60200 ft. would have much larger percentage errors.

Slide rule computation produces the correct significant figures of the answer but it does not locate the decimal point; hence unless some practically infallible method for locating the decimal point is learned, slide rule answers will be of little value. There are several rules for locating the decimal point in slide rule computation. The most important are given below.

*Characteristic Method.* "The characteristic of a whole or mixed number is the number of digits before the decimal point and is positive; the characteristic of a decimal fraction is the number of ciphers immediately after the decimal point and is negative." [1] In accordance with this definition, the characteristics of the numbers 375, 6400, 0.037, and 0.503 are +3, +4, −1, and 0 respectively. It is to

[1] J. E. Thompson, *A Textbook in Algebra*, p. 214. D. Van Nostrand Co., Inc., 1935.

be noted that the characteristic of a number as defined above is not the same as the characteristic of a logarithm.

The characteristic method is used in a number of slide rule manuals. Elaborate sets of rules similar to the following may be found in these manuals.

Rule 1. If the slide projects to the right in multiplication, the characteristic of the product is 1 less than the sum of the characteristics of the two factors.

Rule 2. If the slide projects to the left in multiplication, the characteristic of the product equals the sum of the characteristics of the two factors.

Rule 3. If the slide projects to the left in division, the characteristic of the quotient will equal the characteristic of the dividend minus the characteristic of the divisor.

Rule 4. If the slide projects to the right in division, the characteristic of the quotient will equal that of the dividend minus the characteristic of the divisor plus one.

Rule 5. If there are several factors, add the characteristics of all the factors and deduct 1 for each time the slide projects to the right in carrying out the successive steps of the complete setting.

Rule 6. For combined multiplication and division, combine the rules for multiplication and division given above.

The characteristic method for locating the decimal point is unsatisfactory. It is unwise to use the term *characteristic* as it is used in the above rules. Even though the attention of the learner is called to the fact that the characteristic of a number is not the same as the characteristic of a logarithm, the resulting confusion can hardly be avoided. (A number of manuals use similar rules but use the term "number of places" instead of "characteristic.") Again, a large number of rules are required when this method is used. The rules given above apply only to the C and D scales. Other rules are needed for the A and B scales, and in combined multiplication and division confusion is almost certain. Rules 5 and 6 are hard to apply. To help, some slide rules have indicators with decimal pointers. Even these, however, are of little value and are going gradually out of use.

*Inspection Method.* A number of manuals recommend the following procedure: Disregard the decimal point and find on the rule the significant figures of the answer. Then "after the work of the rule has been done, the decimal point can usually be placed by inspection; for example, through a mental survey of the influence of the involved factors upon the result. Where this is not feasible, a rough arithmetical calculation will serve to properly locate the decimal point."[2]

The slide rule answer to $\frac{40.7 \times 316}{193}$ is 668. The decimal point may be located by the mental calculation $\frac{40 \times 300}{200} = 60$. The answer to the original example is about 60; hence the answer, correctly pointed off, is 66.8.

The above method is fairly satisfactory, and may be used in simple problems. In complicated or unusual problems a large number of errors will result from using this method. The author found that teachers of mathematics in his college classes could not get reliable results for complicated problems by the inspection method. (See page 59.)

*Rough or Approximate Calculation Method.* The "rough arithmetical calculation"[3] often is rather complicated, as may be seen from this example from one of the best slide rule manuals:

$$\frac{5.84 \times 24.8 \times 53 \times 0.314}{51.8 \times 0.058 \times 47.8} = ?$$

The answer 168 is obtained from the slide rule by disregarding the decimal points. The location of the decimal point is found as given below:

$$\begin{aligned}
&(1)\ 5.84 \times 24.8 && = 6 \times 25, \text{ approximately} \\
& && = 150 \\
&(2)\ 150 \times 53 && = 150 \times 50, \text{ approximately} \\
& && = 7500 \\
&(3)\ 7500 \times 0.3 && = 2250 \text{ (approximate numerator)}
\end{aligned}$$

[2] William E. Breckenridge, *The Polyphase Slide Rule*, p. 8. Keuffel and Esser Co. New York, 1925.

[3] Ernst R. Breslich and Charles A. Stone, *The Slide Rule*, p. 13. The University of Chicago Press, Chicago, 1930.

(4) $51.8 \times 0.058 = 52 \times 0.06$, approximately
$= 3.1$

(5) $3 \times 47.8 = 3 \times 50$, approximately
$= 150$ (approximate denominator)

(6) $2250 \div 150 = 15$

(7) Hence the answer is 16.8.

It should be noted that the above example was made for rapid mental calculation. It has the factors 5.84 above and 0.058 below and also 53 above and 51.8 below. Most examples would not be so nicely balanced. In spite of this, the "rough arithmetical calculation" quoted is so long that most people would doubtless rather work the example out with a pencil and not bother with a slide rule.

Better results would have been obtained by writing the example as follows:

$$\frac{6 \times 25 \times 50 \times 0.3}{50 \times 0.06 \times 50}$$

and using cancellation. By cancellation, the approximate result is 15. It is apparent that the answer to the original example cannot differ greatly from this approximate answer. The first three significant figures of the answer obtained from the slide rule were 168, so the correct answer to the example is 16.8.

*Standard Numbers.* The best method for locating the decimal point is the standard number method. To understand this method it is necessary to have a thorough concept of standard numbers.

In scientific literature, numbers like 93,000,000 are often written in the form $9.3 \times 10^7$. Numbers written in the form $9.3 \times 10^7$ are called "standard numbers" or numbers in "standard form." A decimal fraction in standard form has a negative exponent for the factor 10. The decimal fraction 0.0000026 written in standard form is $2.6 \times 10^{-6}$.

A number may be changed to standard form as follows:

1. Write the significant figures of the number and place a decimal point after the first significant figure.

2. Start at the new location of the decimal point and count the number of figures between the new and old locations of the decimal point.

3. Write the factor 10 after the number.

4. Write as an exponent of the factor 10, the number obtained in 2. This exponent is positive if it was obtained by counting in the positive direction and negative if it was obtained by counting in the negative direction.

Exercises A and B are suggested for drill in changing numbers to standard form, and in changing numbers in standard form to ordinary numbers.

A. Change to standard form:

| Number | St. Form | Number | St. Form |
|---|---|---|---|
| 6,000,000,000,000 | ($6. \times 10^{12}$) | 186,000 | ($1.86 \times 10^5$) |
| 0.00000350 | ($3.5 \times 10^{-6}$) | 240,000 | ($2.40 \times 10^5$) |
| 0.0000003228 | ($3.228 \times 10^{-7}$) | 29,160 | ($2.916 \times 10^4$) |

B. Change to ordinary numbers:

| | |
|---|---|
| $6.5 \times 10^{11}$ | (Ans. 650,000,000,000) |
| $3.90 \times 10^{-6}$ | (Ans. 0.00000390) |
| $6.660 \times 10^{-8}$ | (Ans. 0.00000006660) |

As pointed out on page 17, writing a number in standard form removes ambiguity with respect to the number of zeros which are significant. For example, if two zeros of the number 3,800,000 are significant, it should be written $3.800 \times 10^6$; if 3,800,000 is written $3.8 \times 10^6$, this indicates that none of the zeros are significant.

Computation with numbers in standard form furnishes a splendid opportunity for drill with exponents. One of the serious criticisms of the presentation of exponents in most textbooks is that the treatment of exponents is largely confined to literal numbers. The pupil who will make no mistake with the highly artificial $3a^3b^4c^{-2} \times 2a^4b^3c^3$, which is often meaningless to him, will give for the product of $3^2 \times 3^3$ some such answer as $9^5$ or $9^6$. Very few pupils will give the correct answer to such practical problems as $(3.67 \times 10^6) \div (2.43 \times 10^{-4})$ or $(3.92 \times 10^4) \times (6.35 \times 10^7)$. The writer recently heard a number of arithmetic teachers who had studied elementary algebra lament the fact that compound interest tables usually end at 20 years. It never occurred to them that $(1.06)^{20} \times (1.06)^{15} = (1.06)^{35}$, and that by multiplying the amounts in the table—in this case (3.2071 × 2.3965)—they could obtain the value of $(1.06)^{35}$ In this way the table could be extended as far as they desired.

Considerable time should be spent with exponents in connection with arithmetical numbers before going to the rather artificial drills with literal, or general numbers, so common in algebra textbooks.

The following example illustrates the use of standard numbers in computation:

$$(6.00 \times 10^4) \times (8.00 \times 10^6) = (6.00 \times 8.00) \times (10^4 \times 10^6)$$

The answer is:

$$48.0 \times 10^{10} \text{ or } 4.80 \times 10^{11}$$

*A number in standard form has but one digit before the decimal point.* Because of this definition, $48.0 \times 10^{10}$ was changed to $4.80 \times 10^{11}$ so that the answer would be in *standard form.* In scientific handbooks physical constants are not always reduced to *true standard form* and occasionally such numbers as $30.4 \times 10^{-6}$, $495 \times 10^{-5}$, $0.37 \times 10^{-10}$ and $0.098 \times 10^{-6}$ may be found. Such numbers may be said to be in *near standard form.*

The following exercises will furnish additional drill with standard numbers:

1. $3.60 \times 10^6 \times 2.40 \times 10^{-4}$
2. $4.50 \times 10^4 \times 3.60 \times 10^3$
3. $8.50 \times 10^{-6} \times 3.20 \times 10^{-2}$
4. $4.80 \times 10^{-6} \times 3.50 \times 10^4$
5. $3.60 \times 10^6 \div (1.80 \times 10^{-4})$
6. $3.60 \times 10^{12} \div (9.00 \times 10^4)$

Answers:
1. $8.64 \times 10^2$
2. $1.62 \times 10^8$
3. $2.72 \times 10^{-7}$
4. $1.68 \times 10^{-1}$
5. $2.00 \times 10^{10}$
6. $4.00 \times 10^7$

It is easy to teach logarithms to a class expert in the use of standard numbers, since the *characteristic of the logarithm of a number is the same as the exponent of 10 in the standard number form.* This fact is illustrated by the numbers given below:

| Number | Standard Form of Number | Exponent of 10 in Standard Form | Characteristic of Logarithm of Number |
|---|---|---|---|
| 8634.2 | $8.6342 \times 10^3$ | 3 | 3 |
| 0.00362 | $3.62 \times 10^{-3}$ | −3 | −3 |
| 3.462 | $3.462 \times 10^0$ | 0 | 0 |
| 276000 | $2.76 \times 10^5$ | 5 | 5 |

Standard numbers have the following advantages:

1. They are the most concise way of writing numbers, as is evident in examples like $8.76 \times 10^{17}$ and $4.35 \times 10^{-20}$.

2. They may be used to show that zeros are significant, as $8.000 \times 10^6$ shows that three zeros of the number 8,000,000 are significant.

3. They furnish valuable drill in the use of exponents.

4. They furnish valuable drill in the use of directed numbers.

5. They expedite computation in science.

6. They are of considerable value in the teaching of logarithms since the exponent of 10 in the standard number form is the characteristic of the logarithm.

7. They furnish the best method of locating the decimal point in slide rule and other mechanical computation, as is shown below.

8. They are of value in rough mental calculations.

*Standard Number Method.* If a one-digit number is multiplied by another one-digit number, there is no difficulty in telling mentally whether the result will be a number of one or two digits, as $2 \times 3 = 6$ and $3 \times 4 = 12$. The use of standard numbers in locating the decimal point in an answer depends on the above fact.

To find the product of $315{,}000 \times 5240$ on the slide rule change both factors to standard form $(3.15 \times 10^5 \times 5.24 \times 10^3)$. The first three significant figures of the product of $3.15 \times 5.24$ as read on the slide rule are 165. As $3 \times 5 = 15$, the correct answer to $3.15 \times 5.24$ must be 16.5. As $10^5 \times 10^3 = 10^8$, the decimal point in 16.5 must be shifted 8 places in the positive direction to obtain the product of $315{,}000 \times 5240$. The product is 1,650,000,000.

*After a little practice, it is not necessary to write the factors in standard form, or even think of them as standard numbers.* What is the product of $3040 \times 915.0$? The first three figures of the product, as obtained on a slide rule, are 278. The decimal point is located in the following manner:

A. Place the point of your pencil after the 3 in the first factor, 3040., and count the number of digits to the decimal point. You count 3 in the positive direction from 3.

$$\frac{3_|040.}{+3}$$

B. Place the point of your pencil after the first figure, 9, in the second factor, 915., and count the digits to the decimal point. The number is +2.

$$\underset{+2}{9\lfloor 15.}$$

C. The sum of the numbers obtained in A and B is + 5. This may be called the *correction number.* It is the sum of the exponents of 10 when the factors are written in standard form.

D. Since 3 × 9 = 27 (using rounded first digits of each factor), the product of 3.04 × 9.15 is 27.8. If the decimal point is shifted 5 places in the positive direction, using the correction number obtained in C, the number 2,780,000 is obtained. Hence, 3040 × 915.0 = 2,780,000.

What is the product of 3,040,000 × 0.000915? The significant figures of the answer are 278. Steps A, B, and C above are illustrated below in graphic form for the product 3,040,000 × 0.000915.

| Negative Direction − ← | | Positive Direction + → |
|---|---|---|
| A. 3 ⌊040000 (+6) | B. 0.0009⌋ 15 (−4) | C. +6 + (−4) = +2 |

The number +2 obtained in C is the correction number. The product of 3.04 × 9.15 is 27.8. As the *correction number* is +2 the decimal point in 27.8 is shifted 2 places in the positive direction to obtain the product of 3,040,000 × 0.000915. The answer is 2,780.

What is the product of 0.00000304 × 9150? The correction number is −3 since −6 + 3 = −3. The decimal point in 27.8 (3.04 × 9.15) must be shifted 3 places to the negative direction to obtain the product of 0.00000304 × 9150. The answer is 0.0278.

The following examples give drill in locating the decimal point.

If 3.92 × 2.18 = 8.55, find the product of the factors given below:

| Factors | Correction Number | Decimal Point in 8.55 to Be Shifted | Adjusted or Final Answer |
|---|---|---|---|
| 392 × 21.8 | +3 | 3 places to the right | 8550. |
| 0.0392 × 2.18 | −2 | 2 places to the left | 0.0855 |
| 0.0000392 × 21.8 | −4 | 4 places to the left | 0.000855 |
| 392000 × 2180 | +8 | 8 places to the right | 855000000. |
| 0.000392 × 0.00218 | −7 | 7 places to the left | 0.000000855 |

Locating the decimal point by the method given above is as easy as counting and may be taught to the average class in 15 minutes. A considerable amount of drill, however, must follow in order to fix the method.

The following examples illustrate the use of the standard number method in fixing the location of the decimal point in division:

Find the quotient of 86500 ÷ 4150.

The quotient of 8.65 ÷ 4.15 = 2.085 (slide rule result with decimal point located by inspection). The correction number in this case is 1. Hence the decimal point in the above quotient, 2.085, is shifted 1 place to the right, giving 20.85 as the final answer.

The method of obtaining the correction number (+1) in this example is illustrated below:

| Dividend | Divisor | Correction Number |
|---|---|---|
| 8 ˌ 6500. | 4 ˌ 150. | (+4) − (+3) = +1 |
| +4 | +3 | |

Notice that the number obtained from the divisor (+3) is subtracted from the number obtained from the dividend (+4) in order to get the correction number (+1). The +4 and +3 are the exponents of 10 when the numbers are written in standard form $(8.65 \times 10^4) \div (4.15 \times 10^3)$; hence the exponent of the 10 in the divisor is subtracted from the exponent of the 10 in the dividend.

Find the quotient of 0.000865 ÷ 4150. The quotient of 8.65 ÷ 4.15 is 2.085 (slide rule result with the decimal point fixed by inspection).

The correction number, which is −7, is obtained as shown below:

| Dividend | Divisor | Correction Number |
|---|---|---|
| 0.0008 ˌ 65 | 4 ˌ 150. | (−4) − (+3) = −7 |
| −4 | +3 | |

The decimal point in 2.085 must be shifted 7 places in the negative direction (left): 0.000865 ÷ 4150 = 0.0000002085.

If the first digit of the divisor is greater than the first figure of the dividend, two figures are used in the dividend to get the inspection quotient and the correction number.

What is the quotient of 12650 ÷ 0.03067?

12.65 ÷ 3.06 = 4.134

The significant figures, 4134, are obtained from the slide rule. The decimal point in 4.134 is fixed by inspection (12 ÷ 3 = 4).

| Dividend | Divisor | Correction Number |
|---|---|---|
| 12 ˪650. | 0.03˩ 06 | (+3) − (−2) = +5 |
| +3 | −2 | |

Since the correction number is +5, the decimal point must be shifted 5 places in the positive direction (right): 12650 ÷ 0.0306 = 413400.

What is the quotient of 12650 ÷ 0.7467?

The significant figures of the answer obtained from the slide rule are 1696. Since the inspection quotient of 12 ÷ 7 is more than 1 and less than 2, 12.650 ÷ 7.46 = 1.696. The correction number for 12650 ÷ 0.746 is +4, therefore 12650 ÷ 0.746 = 16960.

A number of exercises similar to the following should be given for drill in locating the decimal point in division.

If 4.875 ÷ 2.135 = 2.283, find the quotients in the following table:

| Example | Correction Number | Decimal Point in 2.283 Must Be Shifted | Final or Corrected Answer |
|---|---|---|---|
| 48750 ÷ 2135 | +1 | 1 place to the right | 22.83 |
| 487500 ÷ 0.0002135 | +9 | 9 places to the right | 2283000000 |
| 48.75 ÷ 0.0000002135 | +8 | 8 places to the right | 228300000 |
| 0.0004875 ÷ 213500 | −9 | 9 places to the left | 0.000000002283 |
| 0.004875 ÷ 2135 | −6 | 6 places to the left | 0.000002283 |

The standard number method may be used to locate the decimal point when there are several factors. What is the product of 36800 × 6420 × 0.0215? The first three significant figures of the product of 3.68 × 6.42 × 2.15 are 508, and by inspection 4 × 6 × 2 = 48; so the product is 50.8. Note that each of the inspection factors is rounded to one significant figure. To get the product of 36800 × 6420 × 0.0215, find the correction number as shown below:

| 3 ˪6800. | × 6 ˪420. | × 0.02˩ 15 | (+4) + (+3) + (−2) = +5 |
|---|---|---|---|
| +4 | +3 | −2 | |

Since the correction number is +5, the decimal point in the 50.8, found above, must be shifted 5 places to the right, which will give 5080000. as the desired product.

The standard number method is by far the best method of locating the decimal point in an example like that below.

$$\text{Solve:}\quad \frac{88500 \times 0.0145}{32.5 \times 0.000124}.$$

The first 4 significant figures of the answer obtained from a slide rule are 3185. By inspection $\frac{9 \times 1}{3 \times 1} = 3$, so the answer to $\frac{8.85 \times 1.45}{3.25 \times 1.24} = 3.185$. The correction number is found as shown below. Since the correction number is +5, the decimal point in 3.185 must be shifted 5 places in the positive direction in order to get the answer to the original example. The final answer is 318,500. In actual practice all the work shown below may be done mentally in a few seconds.

Factors above the Line

8 ⌊8500. 0.01⌋ 45

$+4 \qquad -2$

$+4 + (-2) = +2$

Factors below the Line

3 ⌊2.5 0.0001⌋ 24

$+1 \qquad -4$

$+1 + (-4) = -3$

Correction Number

$+2 - (-3) = +5$

*The correction number of a factor squared is twice the correction number of the factor.* The correction number of 2460 is +3; hence the correction number for $(2460)^2$, which equals $2460 \times 2460$, is $(+3) + (+3)$, or +6. Similarly, the correction number for $(0.035)^2$ is $(-2) + (-2)$, or $-4$.

*The correction number of a factor cubed is three times the correction number of the factor.* The correction number of 3500 is +3; hence the correction number of $(3500)^3$, or $3500 \times 3500 \times 3500$, is $(+3) + (+3) + (+3)$, or $+9$. Similarly, the correction number of $(0.025)^3$ is $(-2) + (-2) + (-2)$, or $-6$.

In finding the square root of a number it is easy, in many cases, to locate the decimal point in the root by simple inspection. For example, in finding the square root of 5.14 it is evident that the root is a little more than 2; hence the decimal point of the root, whose significant figures are 227, should be placed after 2, making 2.27. When simple inspection does not indicate the correct position of the decimal point of the root, it is always possible to determine it by dividing the given number into periods of two figures each, beginning at the decimal point, as is done in arithmetic when the square root of a number is obtained by following the formula $(a + b)^2 = a^2 + b(2a + b)$. In this case the number of digits in the integral part of the root is equal to the number of periods to the left of the decimal point in the given number. In the great majority of cases in square root examples, the decimal point is located either in this way or by simple inspection. It is evident that these two methods make no use of the standard number method of locating the decimal point.

In solving complex examples by means of the slide rule, where the operations of multiplication, division, and square root are combined, it is occasionally desirable to locate the decimal point in the factor involving the square root by using the standard number method. This makes it possible to fix the position of the decimal point in the final result of the complex example by applying the standard number method uniformly throughout the work. To do this, it becomes necessary to have a simple rule for finding the correction number in square root. Such a rule is given below. Before stating the rule, the following definitions should be given:

When a number is divided into periods of two figures each, beginning with the decimal point, the first period to the left of the decimal point will be called the *units' period;* the next period to the left of the units' period will be called the *first period;* the next period to the left will be called the *second period;* etc. Similarly, the first period to the right of the units' period will be called the *first period;* the second period to the right will be called the *second period;* etc. In other words, the periods are numbered to the right and left of the *units' period* and *not* to the right and left of the decimal point. This scheme of numbering the periods in the number 16743.0076 is illustrated below:

| Number | 1 | 67 | 43. | 00 | 76 |
|---|---|---|---|---|---|
| Period | Second | First | Units | First | Second |

The rule for finding the correction number of the square root of a given number may now be stated as follows:

*Divide the given number into periods of two figures each, beginning with the decimal point, as is often done in finding square root in arithmetic. Then the* CORRECTION NUMBER *of the square root of the given number is numerically equal to the number of the period in which the first significant figure of the given number is found. This correction number is positive if the period containing the first significant figure lies to the* LEFT *of the* UNITS' PERIOD, *and it is negative if the period containing the first significant figure lies to the* RIGHT *of the* UNITS' PERIOD.

The following examples illustrate the above rule:

*Example* 1. The correction number of the square root of 46,385 (4 63 85.) is +2 because the first significant figure, 4, lies in the second period to the left of the units' period, 85.

*Example* 2. The correction number of the square root of 0.0632 (0.06 32) is −1 because the first significant figure, 6, lies in the first period (.06) to the right of the units' period. In this case, the units' period is 00.

*Example* 3. The correction number of the square root of 0.00003721 (0.00 00 37 21) is −3 because the first significant figure, 3, lies in the third period to the right of the units' period.

The process of finding the square root of any number on the slide rule, entirely aside from the question of the correction number, requires that the given number be divided into periods of two figures each. This separation of the given number into periods enables the operator of the slide rule to know whether he should set the given number on the right half, or on the left half, of the A scale before reading the square root on the D scale. Hence the separation of the given number into periods of two figures each serves the dual purpose of guiding the operator in setting the given number correctly on the A scale and also of giving him quickly the correction number of the square root. A few examples are given below to illustrate the use of the correction number in locating the decimal point in square root.

*Example* 1. Find the square root of 234,000. Dividing the number into periods of two figures each, we get 23 40 00. Since the first period to the left (23) contains two figures, the given number is set on the *right half* of the A scale. Under this number the significant figures of the root, 484, are found on the D scale. Since the correction number of the root is +2, the root itself is 484. $\underset{+2}{(4\,\llcorner 84.)}$

*Example* 2. Find the square root of 52,500. Dividing the number into periods, we get 5 25 00. Since the first period to the left (5) contains only one figure, the given number is set on the *left half* of the A scale. The corresponding reading on the D scale is 229. Since the correction number is +2, the square root is 229. $\underset{+2}{(2\,\llcorner 29.)}$

*Example* 3. Find the square root of 238.56. Dividing the number into periods, we get 2 38.56, which shows that the given number should be set on the left half of the A scale. The corresponding reading on the D scale is 1534. The correction number of the square root is +1; hence the square root is 15.34. $\underset{+1}{(1\,\llcorner 5.34)}$

*Example* 4. Find the square root of 0.00175. Dividing the number into periods, we get 0.00 17 50, which shows that the number should be set on the right half of the A scale. The corresponding reading on the D scale is 418. The correction number of the square root is −2; hence the square root is 0.0418. $\underset{-2}{(0.04\,\lrcorner\,18)}$

*Example* 5. Find the square root of 0.3527. Dividing the number into periods, we get 0.35 27, which shows that the given number should be set on the right half of the A scale. The corresponding reading on the D scale is 59. The correction number of the square root is −1; hence the square root is 0.59. $\underset{-1}{(0.5\,\lrcorner\,9)}$

*Example* 6. Find the square root of 79.53. Dividing the number into periods, we get 79.53, which shows that the given number should be set on the right half of the A scale. The corresponding reading on the D scale is 89. The correction number of the square root is 0; hence the square root is 8.9.

When a number is divided into periods of two figures each, beginning with the decimal point, the correction number of the square root of the given number may be found also in the following way:

Place the point of your pencil after the first *period* of the number and count the number of *periods* to the decimal point. The number found by counting is the correction number. This rule is the same as the one used to find the correction number of any factor except for the fact that the word *period* is substituted for the word *figure*.

In the following examples, the correction number is indicated in the usual way.

Find the square root of the following:

| | | |
|---|---|---|
| 0.00 00 09 | 37 64 27. | 9 26.34 |
| −3 | +2 | +1 |
| 0.00 00 00 86 34 | 3 86 42. | 83.26 |
| −4 | +2 | 0 |

We are now ready to apply the above work to complex examples involving square root with other operations.

In the example below, the correction number of each factor was obtained by using the methods already described. The final correction number is seen to be +3. The significant figures of the final answer, as obtained from the slide rule, are 726. Before applying the correction number to 726, it is necessary to know whether to start after the 7 or after the 72 in using the correction number. This information is obtained by the process of rough approximation, which gives a value of from 6 to 8 depending upon how carefully the mental calculations are made. This value of 6 to 8 shows that we should start after the 7 of 726 to use the final correction number, +3. By doing this, we get a final answer to the original example of 7260.

*Example* 1.

$$\frac{\overset{+6}{(2460)^2} \times \overset{-1}{\sqrt{0.7854}}}{\underset{+2}{\sqrt{54\ 50\ 00}}}$$

Correction Number

$$(+6) + (-1) - (+2) = +3$$

Rough Approximation

$$\frac{(2.5)^2 \times \sqrt{78}}{\sqrt{54}} = 6 \text{ to } 8$$

*Answer:* $7\underset{+3}{\llcorner 260}$

In connection with the rough approximation mentioned above, it is evident that any final estimate between 2 and 9 would give the necessary guidance in using the correction number; hence the various calculations involved in this approximation may be done very roughly.

The significant figures of the final answer to Example 2, as obtained from the slide rule, are 1035. The final correction number is +2. The rough approximation gives a value of about 1, which shows that we should start after the first figure of 1035 when using the correction number, +2. The final answer is 103.5.

*Example 2.*

$$\frac{\overset{+3}{9650} \times \overset{-2}{\sqrt{0.00\ 03\ 46}}}{\underset{+2}{\sqrt{43\ 75\ 00}} \times \underset{-3}{0.00262}}$$

Correction Number

$$+3 + (-2) - (+2) - (-3) = +2$$

Rough Approximation

$$\frac{10 \times \sqrt{3}}{\sqrt{40} \times 3} \text{ or } \frac{10 \times 2}{6 \times 3} = 1$$

*Answer:* $1\underset{+2}{\llcorner 03.5}$

The significant figures of the final answer to Example 3, as found by the slide rule, are 2422. The correction number is +4. In the rough approximation the answer 21 is obtained, which clearly shows that we should start after the *second* figure of the slide rule answer, 2422, when using the correction number, +4. The final answer, therefore, is 242200.

*Example 3.*

$$\frac{\overset{+3}{6780} \times \overset{+1}{32.8} \times \overset{0}{2.14}}{\underset{0}{\sqrt{3.85}}}$$

Correction Number

$$+3 + 1 + 0 - 0 = +4$$

Rough Approximation

$$\frac{7 \times 3 \times 2}{\sqrt{4}} = 21$$

*Answer:* $24 \llcorner 2200.$ (+4)

The decimal point in the final result could have been located in each of the above examples by some form of inspection, but only an expert would be able to do this *mentally* and *always* be sure of the answer. The author has found that high school and college students have no difficulty in placing the decimal point properly when they use the standard number method, but they cannot be depended upon to do this accurately by inspection even in simple examples.

To compare the standard number method for locating the decimal point with the method of "simple inspection," a class of 72 graduate students at Teachers College was given Test B, shown below. These students were asked to locate the decimal point by inspection. They were allowed 9 minutes for the test and most of them took from 7 to 9 minutes. After working the test, the class was given one lesson on the standard number method and was asked to practice this method. Ten days later the test was repeated. More than half the class finished the test in one minute and practically all finished in less than 3 minutes. Test B and an analysis of the results obtained on this test (Table V) are given below.

## TEST B

Locate the decimal point in the following examples:

| Factors | Significant Figures of the Answer | Answer |
|---|---|---|
| (1) 347100 × 0.0002760 | 9541 | ....... |
| (2) 85620 × 432100 | 3722 | ....... |
| (3) 0.0003450 × 0.00004130 | 1425 | ....... |
| (4) 0.00009150 × 369100 | 3377 | ....... |

TABLE V

Analysis of Errors on Test B

| Number of Errors | Inspection Method | Standard Number Method |
|---|---|---|
| Students having 0 errors............... | 3 | 70 |
| Students having 1 error................ | 5 | 0 |
| Students having 2 errors............... | 5 | 1 |
| Students having 3 errors............... | 13 | 1 |
| Students having 4 errors............... | 46 | 0 |

For examples of the type given on Test B, the standard number method is vastly superior to inspection in both speed and accuracy.

### PREVENTING CONFUSION OF TERTIARY GRADUATIONS

On a 10-inch slide rule, the 10 primary graduations on the C and D scales and the 20 primary graduations on the A and B scales are each marked with a digit. On the C and D scales the secondary graduations between the first two primary graduations are each marked with a digit, but all the other secondary graduations are not numbered. None of the tertiary graduations are marked with digits. *There are always ten secondary spaces between the primary graduations on the A, B, C, and D scales.* It is practically as easy to read or to locate a two-digit number on the C or D scales of a slide rule as it is to read inches and tenths of an inch on a foot rule graduated to inches and tenths of an inch. A ten-year-old child, of average intelligence, can easily be taught to read or to locate such a number as

43 on the D scale of a slide rule. A number like 43 is found by locating 4 and then counting over three secondary graduations to the right. If a pupil can read the primary and secondary digits on a 10-inch rule, he can at once read these digits on a 4-inch rule, a 20-inch rule, a Thatcher rule, or practically any other type of slide rule.

The primary graduation marked 2 on the C and D scales may represent any number of the series $2 \times 10^N$. However, in referring to this graduation, it will be called 2. In the same manner the smallest integral value will be used to name each primary graduation as 1 to 10 on the C and D scales[4] and 1 to 10 and 10 to 100 on the A and B scales. Between 1 and 2 on the C and D scales each secondary space is divided into ten tertiary spaces by nine graduations. Each of these tertiary graduations determines or locates a tertiary digit. Between 2 and 4 on the C and D scales and between 1 and 2 and 10 and 20 on the A and B scales there are only five tertiary spaces in each secondary space. The tertiary graduations locate or determine the even tertiary digits, 2, 4, 6, and 8. The odd tertiary digits 1, 3, 5, and 7 must be found by bisecting one of the five tertiary spaces with the hair line of the indicator. That is, a tertiary 3 is halfway between a tertiary 2 and a tertiary 4. From 4 to 10 on the C and D scales and from 2 to 5 and from 20 to 50 on the A and B scales there is only one tertiary graduation. This locates the tertiary digit 5. All the other tertiary figures must be located by trying to estimate fifths of the small space on each side of the middle tertiary 5. Between 5 and 10 and between 50 and 100 on the A and B scales there are no tertiary graduations. To locate or read the tertiary digit, it is necessary to try to divide one of the small secondary spaces into some required ratio.

To locate or "set" 8.76 on the A scale, 8.7 is found and then 6/10 of the 8th secondary space is taken. To locate 5.43 on the A scale, 5.4 is found and then 3/10 of the 5th secondary space must be estimated.

[4] The primary graduation on the C and D scales that is called 10 in this discussion is the last primary graduation on the right-hand end of the C and D scales; this primary division is actually marked 1. Similarly, the primary graduations called 10 to 100 on the A and B scales are graduations on the right half of these scales; the actual markings of these graduations on the A and B scales of most rules are 1, 2, 3, etc., instead of 10, 20, 30, etc.

The third significant figure is determined in four different ways, depending on the part of the slide rule used, and often necessitates difficult visual interpolation.

On account of the difficulty in locating or reading tertiary digits on the slide rule, it is better to commence instruction on the slide rule with two-figure computation.

It was pointed out on page 60 that it is very easy to read or locate the first two significant figures of a number on any slide rule. The first two significant figures of a number are read or located in exactly the same manner on any part of the slide rule.

After the pupil is familiar with two-figure computation on the slide rule, he may gradually be introduced to three-figure computation. This procedure speeds up slide rule instruction, makes early slide rule work more interesting, and greatly reduces the number of errors caused by confusing tertiary slide rule graduations.

## PREVENTING ERRORS IN ESTIMATING THE THIRD FIGURE

When using sections of the slide rule where there are no tertiary divisions, or only one such division, even an expert cannot always estimate the third figure correctly. A student needs a considerable amount of careful practice before he can read the third figure with a fair degree of accuracy.

To develop skill in reading the third figure the student should find the square roots of numbers that can be exactly located on the A scale, such as 1.42, 2.05, 4.65, 12.4, 26.5, 56.0, etc. In finding the square roots of such numbers practically all the errors will be in reading the answers. The student should not try at the start to estimate both the number and the square root. The slide rule answers should be checked with a square root table. The student should try to read the answer correct to three significant figures. In checking slide rule answers with a table, the answers in the table should be rounded to three significant figures. A number like 3.468 should be rounded to 3.47. With practice it is possible to get the third figure correct seven or eight times out of ten. If the student does not get the third figure correct the first time he should repeat the operation and see why he failed. Exercises in reading sines and tangents and checking with a table furnish similar drill in reading the third significant figure. In reading square roots, sines, and tangents, it is not necessary to manipulate the slide rule by moving the slide.

If the student obtains answers which are always slightly too large, or always slightly too small, he should check the hairline of the indicator of the rule by getting the square roots of 64, 25, 16, 9, and 4. If the answers are not exactly 8, 5, 4, 3, and 2, the indicator should be adjusted.

Additional drill in estimating the third significant figure may be obtained by reading the answers to problems in multiplication and division in which the factors may be exactly located by secondary or tertiary graduations.

In many cases the error in the third figure of the answer is caused by careless settings. For example, suppose we wish to divide 875 by 725. To do this we set 875 on D and place 725 on C over it. If the setting is correct, the quotient, 1.206 (or 1.207) is obtained. In making the above setting, the hairline of the indicator is placed on 875 (D) and then 725 (C) is placed under the hairline. If the setting is carefully made, the graduations for 725 and 875 will form one continuous line, when the hairline of the indicator is shifted to the left as shown in A. If the graduations are out of alignment, as in B or C, there will be an error in the answer. In B and C the hairline of the indicator is probably between the 875 and the 725 graduations, thus filling up the gap.

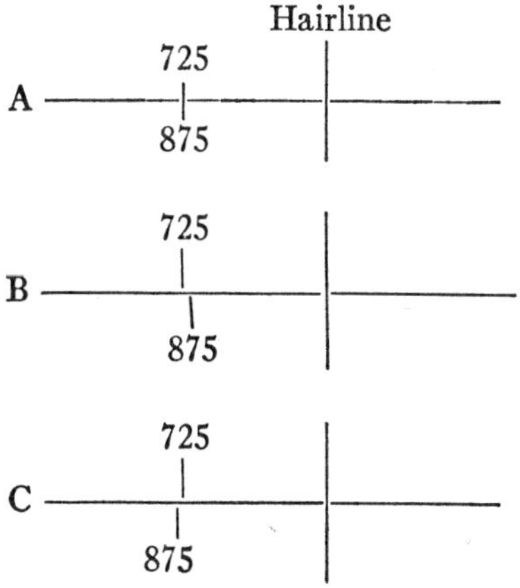

*A plus or minus tolerance of around 1 in 1000 is usually allowed for slide rule solutions.* If the correct answer to an example is 9.68, full credit is given for the answers 9.67 or 9.69. If the correct answer is 13.76, full credit is given for 13.75 or 13.77. If a student has trouble in getting answers within the tolerance allowed, he should practice with such exercises as those suggested on pages 62 and 63. It should be noted that the slide rule tolerance of 1 in 1000 is smaller than that allowed on any new commercial weighing or measuring instrument, with the exception of a steel tape. Even a tolerance of 1 in 500 is smaller than that allowed on all but four items. (see page 30).

## PREVENTING ZERO ERRORS

All the primary or main graduations, such as 1, 2, 3, etc., should be read 1.00, 2.00, 3.00 when working with three-digit data. The zeros should be written to show that the answers are correct to 3 significant figures. Students should be given considerable drill in reading and locating the numbers 1.01 to 1.09, 2.01 to 2.09, 3.01 to 3.09, etc., on the C and D scales. They should also read and locate these numbers on the A and B scales where this is possible. They should also be taught to read and locate on the C and D scales such numbers as 1.005, 2.005, and 3.005.

Following the above drill there should be additional drill in reading or setting zeros, with examples similar to those below.

Find the following square roots to three figures. Watch for the zeros.

1. $\sqrt{4.25}$ (2.06)
2. $\sqrt{4.20}$ (2.05)
3. $\sqrt{9.3}$ (3.04)
4. $\sqrt{9.5}$ (3.08)
5. $\sqrt{2.10}$ (1.45)
6. $\sqrt{2.05}$ (1.43)
7. $\sqrt{2.07}$ (1.44)
8. $\sqrt{10.0}$ (3.16)
9. $\sqrt{9.50}$ (3.08)

Solve the following on the slide rule:

10. $132 \div 3.25$ (40.5)
11. $700 \div 66.0$ (10.6)
12. $20.4 \times 15.1$ (308)
13. $9.05 \times 12.05$ (109)

Solve the following. Watch for zeros in your answers.

14. $\frac{\times}{230} = \frac{550}{620}$ (204)
15. $\frac{\times}{13.0} = \frac{15.0}{19.4}$ (1.005)
16. $\frac{1.87}{1.50} = \frac{\times}{4.05}$ (5.05)
17. $\frac{\times}{21.2} = \frac{46.5}{32.0}$ (30.8)

In subsequent work on the slide rule, frequent drill should be given with problems having zero for a secondary or tertiary digit in the factors or in the answers. Students must be made and kept "zero conscious" in order to avoid serious zero errors. If this is done, zero errors may be largely eliminated.

# CHAPTER VIII

## TESTING THE EFFICIENCY OF THE METHODS OF PREVENTING SLIDE RULE ERRORS

IN ORDER to test the efficiency of the methods described in Chapter VII, Test A was given to an experimental group of 150 students who had been instructed by the new methods. These students were given the same amount of instruction as the 660 students described in Chapter VI, who had previously taken the same test. Both groups of students received class instruction for six fifty-minute periods and were asked to practice two hours for each class period. The slide rules used by the students in both groups were of the cheaper $1.00 to $1.25 type.

Fifty-four of the experimental group were students at State Teachers College, Trenton, N. J., 59 were students at Hunter College, and 37 were students at Teachers College, Columbia University.

Test A was also given to a class of 22 high school pupils and a volunteer class of 14 eighth grade pupils. These two groups had the same instruction as the 150 college students.

A comparison of the two groups with respect to age and initial ability is given on pages 78 to 80. The methods of instruction used with the two groups are given below.

| *Methods of Instruction Used with the Original 660 Students* | *Methods of Instruction Used with the 150 Students* |
|---|---|
| 1. In the first lesson the students were taught to read and locate three-figure numbers on the slide rule and to multiply one-, two- and three-figure numbers. No previous instruction was | 1. Students were taught to read *primary* and *secondary* slide rule graduations only, and to find the first two figures of the square root of all numbers from 1 to 100. Students were given |

| *Methods of Instruction Used with the Original 660 Students* | *Methods of Instruction Used with the 150 Students* |
|---|---|
| given in approximate computation. | work in approximate computation and standard numbers before studying the slide rule. |
| 2. Students were taught to locate the decimal point in answers by simple inspection, or by rough arithmetical calculation as suggested by William E. Breckenridge[1] and Ernst R. Breslich.[2] | 2. Students were taught to use standard numbers or the standard number method (pp. 49–60) and were encouraged to use this method in all cases except those where simple inspection could be used. |
| 3. No special drill was given with examples involving numbers having zero as a secondary or tertiary figure. Students were taught to read or set secondary zeros that were met in problems, but no special emphasis was placed on such work. | 3. Students were given extensive drill in reading and locating numbers like 101, 105, 201, 205, 208, 302, 405, etc. They were also given numerous examples with zero as a secondary figure in one of the factors or in the answer in order to make them zero conscious. |
| 4. No special drill was given to help students estimate the tertiary slide rule figures. Students spent time on drill exercises and other slide rule problems. | 4. Students were asked to practice estimating the tertiary figure of slide rule answers by getting the square root of numbers that could be exactly located on the A scale, such as 2.45, 4.65, 13.6, 22.5, and by using factors in multiplication and division that could be exactly set, such as 248 × 326 or 328 × 465. |
| 5. Students were taught to read primary, secondary, and tertiary | 5. Students were taught to read and set primary and secondary |

[1] William E. Breckenridge, *The Log Duplex Slide Rule*, p. 8. Keuffel and Esser Co. New York, 1926.

[2] E. R. Breslich and C. A. Stone, *The Slide Rule*, pp. 11 and 13. The University of Chicago Press, Chicago, 1929.

| *Methods of Instruction Used with the Original 660 Students* | *Methods of Instruction Used with the 150 Students* |
|---|---|
| graduations in the first lesson. They then spent their time on drill exercises involving two- and three-figure numbers. | figures in the first lesson. After they had acquired some skill in using the slide rule for two-figure computation, they were then taught in the third lesson to read tertiary figures with the four types of tertiary graduations. (See pp. 60–62.) |

Table VI gives an analysis of the errors that were made by the four groups of students who took Test A. Only types A, B, C, and D are given.

With respect to decimal point errors, Table VI shows that the original 660 college students averaged 1.44 errors per student, whereas the experimental group of 150 college students averaged .38 errors per student. In other words, the 660 students made about *four times* as many decimal point errors as the 150 students. Making a similar comparison with respect to the errors in estimating the third significant figure, we find that the 660 students averaged about *twice* as many errors per student as the 150 students. With respect to the zero errors, the 660 students made about *three times* as many errors as the 150 students. The results with respect to these three errors therefore seem to be outstandingly in favor of the 150-student group. With respect to errors in reading the tertiary graduation, however, the superiority of one group over the other is not so evident, the 660 group averaging .59 errors per student and the 150 group averaging .43 errors. It is evident, however, that the 150 group did somewhat better than the 660 group. It is quite possible, with respect to the errors in reading the tertiary graduations, that another element, other than the methods of instruction, may have influenced this result; namely, the fact that some of the slide rules used by the two groups were of the cheaper variety upon which the markings of the tertiary graduations are sometimes not as clean cut, and not as clearly distinguishable from the secondary graduations, as they are on the better slide rules. The standard method of distinguishing tertiary graduations from secondary graduations is to make the tertiary graduations a little shorter than the secondary. If the difference in these two lengths is very small, and if the engraving is not clean cut, the eye does not always distinguish easily between the two types of markings. For this reason, errors in reading the tertiary graduations may sometimes occur which might have been avoided if the rules had been more carefully engraved.

TABLE VI

Errors Made on Test A by Four Different Groups of Students

| | 660 College Students | | 150 College Students | | 22 High School Pupils | | 14 Eighth Grade Pupils | |
|---|---|---|---|---|---|---|---|---|
| Type of Error | Total No. of Errors | Errors per Student | Total No. of Errors | Errors per Student | Total No. of Errors | Errors per Student | Total No. of Errors | Errors per Student |
| Decimal point errors...... | 952 | 1.44 | 57 | 0.38 | 16 | 0.73 | 63 | 4.50 |
| Errors in estimating third significant figure...... | 723 | 1.10 | 92 | 0.61 | 9 | 0.41 | 11 | 0.79 |
| Zero errors... | 599 | 0.91 | 45 | 0.30 | 8 | 0.36 | 10 | 0.71 |
| Errors in reading tertiary graduations | 389 | 0.59 | 65 | 0.43 | 5 | 0.23 | 2 | 0.14 |
| Total.... | 2,663 | 4.04 | 259 | 1.72 | 38 | 1.73 | 86 | 6.14 |

*Note:* The four types of errors mentioned above are described on pp. 38 to 41.

No attempt will be made to draw conclusions with respect to the performances of the 22 high school pupils or the 14 eighth grade pupils in comparison with the college students since these two groups each involve so few numbers. It is evident, however, from Table VI that the 22 high school pupils made a much better record on all four types of errors than the 660 college students. It is also evident from the table that the 14 eighth grade pupils did better than the 660 college students with respect to all errors except those relating to the decimal point. With reference to the decimal point errors, these 14 eighth grade pupils averaged 4.50 errors per pupil in comparison with 1.44 errors per student for the 660 college students and .38 errors per student for the 150 college students.

An additional fact should now be brought out concerning the decimal point errors made by the 660 college students and those made by the 150 college students. It has already been stated that the 150 college students were taught to locate the decimal point by the aid of standard numbers, whereas the 660 college students did not follow this method. Concerning the method used with the 150 students, it is important to point out that part of these students were taught one variety of the standard number procedure while the rest were taught another variety of it. To be specific, 59 of the 150 students were taught to locate the decimal point by using standard numbers in the manner described on pages 46 to 48. By this method the student writes each factor in standard form as illustrated on page 48, and uses the laws of exponents. The answer is found in standard form and then changed back to an ordinary number. The remaining 91 of the 150 students were given the *standard number method* for locating the decimal point; this method, described on pages 49-59, is based on standard numbers but so simplified that no pencil work is required. The student works only with the exponents, without stopping to write the factors in the full standard number form; the advantage of this procedure is that it is faster than that of standard numbers and reduces the chances of error.

Table VII shows the results of these two procedures for locating the decimal point, used by the 59 and 91 students from the experimental group of 150 students, and gives for comparison the results obtained by inspection with the original 660 students.

It is evident from Table VII that the original 660 students made twice as many decimal point errors per paper as the 59 students using standard numbers and about 9 times as many errors per paper as the 91 students using the *standard number method.* It is this *standard number method* for locating the decimal point that is advocated by the author in this study.

We will now proceed to make a more technical comparison of the

TABLE VII

Number of Errors Made by Three Groups of Students in Locating the Decimal Point

| | 59 Students Using Standard Numbers | 91 Students Using Standard Numbers Method | Original 660 Students |
|---|---|---|---|
| Total number of decimal errors.................. | 42 | 15 | 952 |
| Mean number of decimal errors per paper......... | 0.71 | 0.165 | 1.44 |

original 660 college students with the latter group of 150 college students. The distributions of the scores of these groups of students on each of the four chief errors are given in Table VIII. The difference of the means, the standard error of the difference, the ratio of the difference divided by its standard error, and the value of $\chi^2$ (chi square) were computed for each of the four chief types of errors and appear at the bottom of the table.

It may be seen from Table VIII that the ratios $\frac{D}{\sigma_D}$ for the decimal point errors, the errors in estimating the third significant figure, and the zero errors are 11.78, 5.27, and 10.34 respectively, each of these ratios being considerably greater than 3.00. Hence, with respect to these three errors, the difference between the performance of the 150-student group and that of the 660-student group should be considered statistically significant. All the differences are in favor of the experimental group.

With respect to the errors in reading the tertiary graduations, the ratio $\frac{D}{\sigma_D}$ is 2.67, as shown in Table VIII. A ratio of this size is usually considered significant.

These conclusions are corroborated by the values of $\chi^2$ in each of the four cases. The obtained $\chi^2$ is greater than the value of $\chi^2_{.01}$ in each case, with the exception of the tertiary graduations. The ex-

TABLE VIII

Summary of Results on Test A

| Errors per Student | Type of Error: Decimal Point (A) N = 660 | Decimal Point (A) N = 150 | Third Digit (B) N = 660 | Third Digit (B) N = 150 | Zero (C) N = 660 | Zero (C) N = 150 | Tertiary Graduation (D) N = 660 | Tertiary Graduation (D) N = 150 |
|---|---|---|---|---|---|---|---|---|
| 0. . . . . . . . . . . | 250 | 111 | 272 | 92 | 267 | 114 | 362 | 94 |
| 1. . . . . . . . . . . | 152 | 27 | 198 | 37 | 235 | 27 | 230 | 49 |
| 2. . . . . . . . . . . | 109 | 8 | 105 | 12 | 120 | 9 | 52 | 5 |
| 3. . . . . . . . . . . | 65 | 2 | 44 | 6 | 28 | | 10 | 2 |
| 4. . . . . . . . . . . | 40 | 2 | 30 | 2 | 10 | | 5 | |
| 5. . . . . . . . . . . | 24 | | 5 | 1 | | | 1 | |
| 6. . . . . . . . . . . | 9 | | 4 | | | | | |
| 7. . . . . . . . . . . | 2 | | 2 | | | | | |
| 8. . . . . . . . . . . | 1 | | | | | | | |
| 9. . . . . . . . . . . | 2 | | | | | | | |
| 10. . . . . . . . . . . | 0 | | | | | | | |
| 11. . . . . . . . . . . | 0 | | | | | | | |
| 12. . . . . . . . . . . | 0 | | | | | | | |
| 13. . . . . . . . . . . | 1 | | | | | | | |
| Mean. . . . . . . . | 1.44 | 0.38 | 1.10 | 0.61 | 0.91 | 0.30 | 0.59 | 0.43 |
| $S^2$ * . . . . . . . . . | 2.813 | 0.582 | 1.620 | 0.931 | 0.884 | 0.330 | 0.612 | 0.392 |
| S. . . . . . . . . . . | 1.68 | 0.76 | 1.27 | 0.96 | 0.94 | 0.57 | 0.78 | 0.63 |
| $S_m$. . . . . . . . . . | 0.065 | 0.062 | 0.049 | 0.078 | 0.037 | 0.046 | 0.030 | 0.051 |
| Difference. . . . | 1.06 | | 0.49 | | 0.61 | | 0.16 | |
| $\sigma_D$. . . . . . . . . . | 0.09 | | 0.093 | | 0.059 | | 0.060 | |
| $D \div \sigma_D$. . . . . | 11.78 | | 5.27 | | 10.34 | | 2.67 | |
| $\chi^2$ . . . . . . . . . | 53.57 (3)† | | 18.75 (4) | | 53.04 (1) | | 4.42 (2) | |
| $\chi^2_{.01}$ . . . . . . . . | 11.34 (3) | | 13.28 (4) | | 6.64 (1) | | 9.21 (2) | |

* In this study S is used for the sample estimate of the paramater $\sigma$.

† Figures in parentheses indicate degrees of freedom.

treme improbability that the experimental scores are a sampling variation of a population of which the scores of the 660 group are another random sample enables us to say with considerable confi-

TABLE IX

Distribution of the Number of Wrong Answers on Test A

| Number of Wrong Answers | Original 660 Students | | 150 Experimental Students | |
|---|---|---|---|---|
| | Number | Per Cent | Number | Per Cent |
| 0 | 27 | 4.1 | 39 | 26.0 |
| 1 | 105 | 15.9 | 40 | 26.7 |
| 2 | 97 | 14.7 | 26 | 17.3 |
| 3 | 96 | 14.6 | 19 | 12.7 |
| 4 | 96 | 14.5 | 13 | 8.7 |
| 5 | 72 | 10.9 | 7 | 4.7 |
| 6 | 70 | 10.6 | 5 | 3.3 |
| 7 | 52 | 7.8 | 0 | 0.0 |
| 8 | 26 | 3.9 | 1 | 0.6 |
| 9 | 13 | 1.9 | | |
| 10–14 | 8 | 1.0 | | |
| Mean | 3.82 | | 1.83 | |
| $S^2$ | 5.53 | | 2.93 | |
| S | 2.35 | | 1.71 | |
| $S_m$ | 0.092 | | 0.140 | |
| Difference | 1.99 (In favor experimental group) | | | |
| $\sigma_D$ | 0.651 | | | |
| $D \div \sigma_D$ | 3.06 | | | |

*Note:* The distribution of the number of errors by types for each of the 15 examples on Test A is given in Tables XI and XII.

dence that the divergencies between the sets of scores for the first three errors are significant.

The difference between the means of two Poisson distributions is not normally distributed in small samples, but the distribution of that difference approaches the normal as the number of cases increases. With samples as large as those in this study it is safe to refer the difference to a normal probability table.

The graphs of the pairs of distributions for each of the four chief errors are shown on page 75. A Poisson distribution with the same mean as the distribution obtained in the study has been plotted along with each distribution. The $\chi^2$ showing the goodness of fit, together with $\chi^2_{.01}$ for each distribution when compared with the Poisson distribution, is also given.

The graphs on page 75 show that six of the eight distributions are in close agreement with the corresponding theoretical Poisson distributions. This agreement is also shown by the fact that in most of these distributions the mean is approximately equal to the variance.

TABLE X

Errors of Highest 90% and Lowest 10% of the Students on Test A

| | Original 660 Students | Experimental 150 Students |
|---|---|---|
| A. Number of perfect papers.................. | 27 | 39 |
| B. Total number of errors made by lowest 10%... | 623 | 81 |
| C. Per cent of total errors made by lowest 10%... | 22% | 28% |
| D. Average errors per paper by lowest 10%...... | 9.5 | 5.4 |
| E. Average errors per paper by highest 90%..... | 3.5 | 1.4 |

As stated on page 38, it was possible for a student to make several errors in working the same example. It was therefore possible for a student to have a total of more than 15 errors on Test A. It was not possible, however, to have more than 15 wrong answers on the 15 problems. The distribution of wrong answers, together with the various computed statistics, is given in Table IX. Here again the ratio $\frac{D}{\sigma_D}$ is large enough to indicate that the scores of the experimental group were significantly superior to those of the original group.

In any class in mathematics there is always a small per cent of

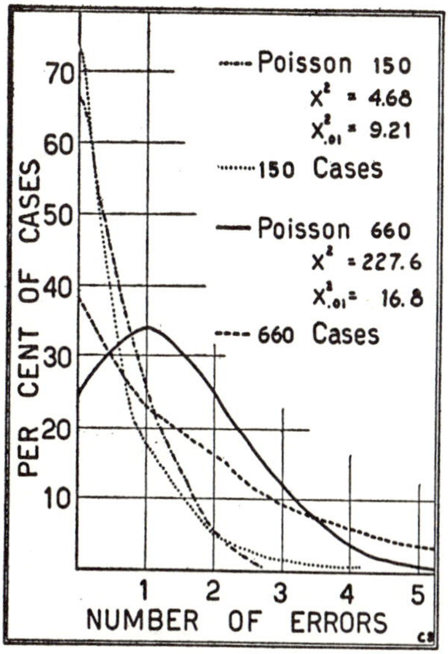


DECIMAL ERRORS

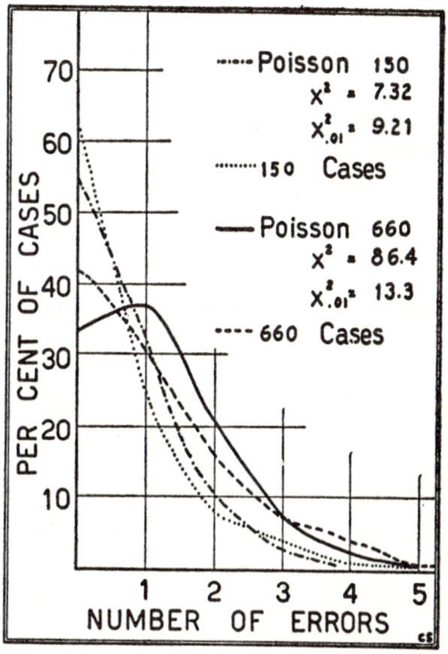


ERRORS IN THIRD FIGURE

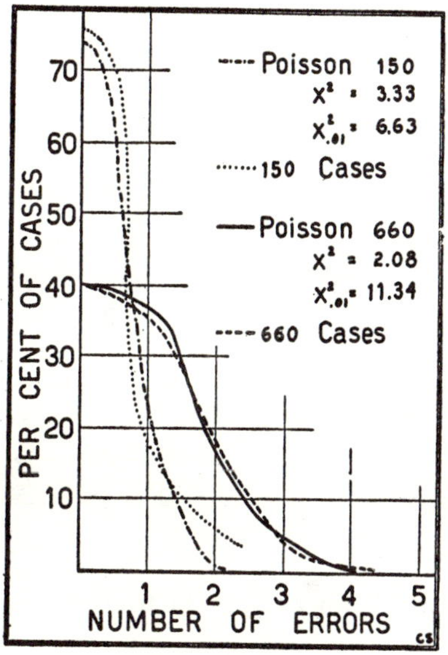


ZERO ERRORS

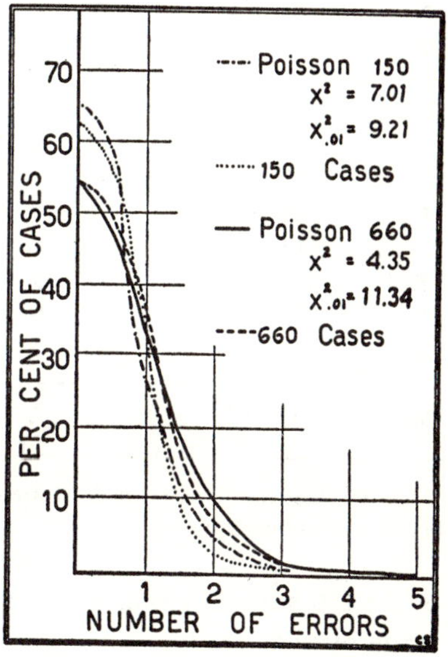


ERRORS, TERTIARY GRADUATIONS

students who either lack the manual dexterity to operate a slide rule successfully or have defects of eyesight that cause faulty readings of the graduations. These students make many mistakes and this greatly reduces the average performance of the entire class. Table X shows how the lowest 10% of each group performed on Test A, in comparison with the highest 90% of each group. From this table it is seen that the lowest 10% of the 660-student group contributed 22% of the total errors made by that group, while the lowest 10%

TABLE XI

Errors of the 660 Students, Arranged According to Examples

| Number of Example | Types of Errors | | | | | | | | | | | | | | Total |
|---|---|---|---|---|---|---|---|---|---|---|---|---|---|---|---|
| | A* | B | C | D | E | F | G | H | I | J | K | L | M | N | |
| 1 | 67 | 35 | 199 | 87 | | 1 | | 4 | | | 1 | 1 | 2 | | 397 |
| 2 | 18 | 9 | 3 | 38 | | 1 | | 3 | | 1 | | 2 | | | 75 |
| 3 | 62 | 50 | 112 | 36 | | 3 | | | | | | | | | 263 |
| 4 | 40 | 41 | 9 | 18 | | 7 | | | | | | | | | 115 |
| 5 | 37 | 24 | 7 | 2 | 2 | 2 | | | 1 | | | | | | 75 |
| 6 | 169 | 38 | 56 | 8 | 11 | 6 | 25 | 1 | | 1 | | | | | 315 |
| 7 | 16 | 43 | 31 | 27 | | | | | | 1 | | | | | 118 |
| 8 | 29 | 40 | 6 | 7 | | 7 | | 2 | 1 | | | | | | 92 |
| 9 | 5 | 44 | | 22 | 6 | 9 | 12 | | | 1 | 1 | | | | 100 |
| 10 | 21 | 41 | 161 | 36 | | | | | | | | | 2 | | 261 |
| 11 | 18 | 69 | 1 | 2 | 10 | 2 | | | | 1 | | 1 | | | 104 |
| 12 | 13 | 74 | 14 | | | 4 | | | | | 3 | | | | 108 |
| 13 | 107 | 59 | | 12 | 9 | 2 | | 2 | 3 | 1 | | | 1 | | 196 |
| 14 | 159 | 82 | | 46 | 9 | 2 | | | 3 | | | | | 2 | 303 |
| 15 | 191 | 74 | | 48 | 5 | 5 | | 3 | | 1 | 1 | 2 | | 1 | 331 |
| Total | 952 | 723 | 599 | 389 | 52 | 51 | 37 | 15 | 8 | 7 | 6 | 6 | 5 | 3 | 2,853 |

* A, B, C, etc., are the types illustrated on page 39. The columns show the number of errors of each type.

of the 150-student group were responsible for 28% of the total errors made by that group.

Comparing the lowest 10% in each case with the highest 90%, it is seen for the 660-student group that the lowest 10% averaged 9.5 errors per paper, while the highest 90% averaged only 3.5 errors per paper. Similarly, for the 150-student group, the lowest 10% averaged 5.4 errors per paper. It is evident that the lowest 10% in each case did very poorly in comparison with the 90%. In general this lowest 10% should be discouraged from continuing work with the slide rule, particularly when it is evident that the fault is due to poor eyesight or lack of manual dexterity.

TABLE XII

Errors of the 150 Students, Arranged According to Examples

| Number of Example | Types of Errors | | | | | | | | | | | | | | Total |
|---|---|---|---|---|---|---|---|---|---|---|---|---|---|---|---|
| | A* | B | C | D | E | F | G | H | I | J | K | L | M | N | |
| 1....... | 2 | 2 | 21 | 12 | | | | 1 | | | | | | | 38 |
| 2....... | | 2 | | 5 | | 1 | | 1 | | | | 1 | 1 | | 11 |
| 3....... | 1 | 7 | 9 | 3 | | | | | | | | | | | 20 |
| 4....... | 2 | 2 | | 1 | | 1 | | | | | | | | | 6 |
| 5....... | 1 | 2 | | | | 3 | 4 | | | | | | | | 10 |
| 6....... | 13 | 2 | 1 | 1 | 3 | 2 | | | | | | | | | 22 |
| 7....... | | 3 | 4 | 6 | | | | | | | | | | | 13 |
| 8....... | 1 | 4 | | 1 | | 1 | | 1 | | | | | | | 8 |
| 9....... | | 3 | | 2 | 1 | | 2 | | | 1 | | | | | 9 |
| 10....... | | 2 | 8 | 6 | | | | | | | | | | | 16 |
| 11....... | | 9 | | | 2 | 1 | | | | | | | | | 12 |
| 12....... | | 14 | 2 | 1 | | | | | | | 1 | | | | 18 |
| 13....... | 6 | 12 | | 3 | 2 | 1 | | | 1 | | | | | | 25 |
| 14....... | 18 | 17 | | 10 | 1 | | | | 2 | | | | | 1 | 49 |
| 15....... | 13 | 11 | | 14 | 2 | 3 | | 2 | | 1 | | 1 | 1 | | 48 |
| Total... | 57 | 92 | 45 | 65 | 11 | 13 | 6 | 5 | 3 | 2 | 1 | 2 | 2 | 1 | 305 |

Tables XI and XII show the number of errors of each type, A to N, for the two groups of 660 and 150 students. Table XI shows that the 660 students made 67 decimal errors on Problem 1. They made

* A, B, C, etc., are the types as illustrated on page 39. The columns show the numbers of errors of each type.

35 errors in estimating the third digit on problem 1 and 199 zero errors on the same problem.

*Effect of Age on Learning to Use the Slide Rule.* Table XIII lists the schools attended by the students to whom Test A was administered.

TABLE XIII

Institutions Attended by Students Taking Test A

| Institution | 660 Students | 150 Students |
|---|---|---|
| Teachers College, Columbia University.......... | 554 | 37 |
| State Teachers College, Trenton, N. J............ | 73 | 54 |
| Hunter College, New York City................ | 33 | 59 |

Most of the 554 students from the various mathematic classes in Teachers College, Columbia University, were graduate students. The students taking Test A at Trenton and Hunter College were mostly sophomores. The average student in the 660 group was older than the average student in the 150 group. Was this difference in age sufficient to give the younger students a significant advantage over the older students?

F. H. Achard,[2] of the Educational Bureau of the Consolidated Edison Company of New York, investigated the effect of age on the ability to use the slide rule. In 1936 he gave a course on the slide rule to 172 of the Company's employees. At the end of the course a rather difficult examination was given, with the results shown in Table XIV.

Most of the students in the 660 group who took Test A were in the second, third, and fourth age groups of Table XIV, while most of the 150 group were in the first, second, and third age groups of Table XIV.

Achard's study shows that a slightly lower score accompanies the greater age (above 36), but the differences between the means of the 21–25, 26–30, 31–35, and 36–40 age groups are not large. As

[2] F. H. Achard, "Teaching an Old Dog New Tricks." *Trained Men*, Autumn Number, 1936, pp. 32 and 33.

the few older students (above 36) among the group of 660 were experienced teachers of mathematics and in many cases superior students, it seems reasonable to assume that the greater age of these students was largely compensated for by their greater mathematical maturity.

TABLE XIV

Mean Scores Made on Slide Rule Test by Various Age Groups

| Age Group | 16–20 | 21–25 | 26–30 | 31–35 | 36–40 | 41–45 | 45–Up | All Ages |
|---|---|---|---|---|---|---|---|---|
| Number of employees.. | 10 | 46 | 49 | 35 | 15 | 4 | 3 | 172 |
| Average scores........ | 77.5 | 81.1 | 83.0 | 82.4 | 78.7 | 71.7 | 74.7 | 81.2 |

The above test was given to Consolidated Edison employees and classified by age groups.

The original 660 students included 73 students from the State Teachers College at Trenton. The 150 students included 54 students from the same college. These two groups of Trenton students were of approximately the same age level, were taught by the same teacher, and it was possible to obtain considerably more information concerning their previous training and mathematical ability than could be obtained for the other students.

The grades of the Trenton students in high school algebra I and plane geometry and their grades on the local college entrance examination were obtained and tabulated. The local college entrance examinations in mathematics are of about the same degree of difficulty from year to year. Since all students who took Test A were required to take algebra I and plane geometry in high school, it was possible to obtain averages for all the students in these two subjects. The local college entrance examination is a general examination including arithmetic, algebra, and rather easy geometric concepts.

It was not possible to secure the above data for all of the State Teachers College students since some of these students transferred from other institutions and did not take the entrance examination. It was therefore necessary to eliminate 8 students from the group of 73 on the original test and 5 students from the latter group of 54. This reduced the two groups to 65 students and 49 students respectively.

The means and standard deviations of these students in algebra I, geometry, and the college entrance examinations are shown in Table XV. The difference of the means, the standard error of the difference, and the ratio of the difference divided by its standard error were determined for each of the three subjects given in Table XV, and are shown in Table XVI.

TABLE XV

Means and Standard Deviations for the Groups of 65 and 49 State Teachers College Students in Algebra, Geometry, and the College Entrance Examinations in Mathematics

| Group | Algebra | | Geometry | | College Entrance Examination in Mathematics | |
|---|---|---|---|---|---|---|
| | M | S | M | S | M | S |
| 65 students from 660 group... | 86.53 | 7.41 | 84.37 | 8.20 | 86.34 | 11.26 |
| 49 students from 150 group... | 85.00 | 7.18 | 81.76 | 10.03 | 82.43 | 12.19 |

TABLE XVI

Test of the Significance of the Difference of the Means of the Groups of 65 and 49 State Teachers College Students on the Subjects in Table XV

| Subject | Difference of Means | Standard Error of Difference | Ratio $\frac{D}{\sigma_D}$ |
|---|---|---|---|
| Algebra.................... | 1.53 | 1.40 | 1.09 |
| Geometry.................. | 2.61 | 1.72 | 1.51 |
| College entrance examination in mathematics .......... | 3.91 | 2.86 | 1.37 |

$\sigma_{(M_1 - M_2)}$ is computed as $\sqrt{\frac{N_1S_1^2 + N_2S_2^2}{N_1 + N_2 - 2} \times \frac{N_1 + N_2}{N_1N_2}}$

The difference of the means is in favor of the original group of 65 State Teachers College students in all three subjects.

The above ratios $\frac{D}{\sigma_D}$ do not indicate a significant difference between the two groups in any of the three subjects. However, whatever difference there might have been was in favor of the 65-student group. In other words the 65-student group on the whole was somewhat superior to the 49-student group, as indicated by the data in Table XV.

The difference of the means, the standard error of the difference, and the ratio of the difference to its standard error were computed for the two groups (65 and 49 students) on the four leading types of errors. These are shown in Table XVII. It should be noted that while the difference of the means with respect to mathematical preparation is in favor of the original group in all three cases in Table XVI, *the difference of the means with respect to slide rule errors is in favor of the experimental group in all four cases given in Table XVII.*

TABLE XVII

Test of the Significance of the Difference of the Means of the Groups of 65 and 49 State Teachers College Students on the Four Chief Types of Errors

| Type of Error | Difference of Means | $\sigma_d$ | Ratio $\frac{D}{\sigma_D}$ |
|---|---|---|---|
| A (Decimal) | 0.900 | 0.223 | 4.09 |
| B (Third digit) | 0.633 | 0.197 | 3.11 |
| C (Zero) | 0.561 | 0.131 | 4.28 |
| D (Tertiary graduation) | 0.075 | 0.291 | 0.82 |

Table XVII corroborates Table VIII with the exception of the fourth case (D). The first three ratios in Table VIII indicate significance (11.78, 5.27, and 10.34), while the fourth ratio (2.67) is hardly large enough to indicate definite significance but would usually be considered significant.

The first three ratios in Table XVII are all so large that for even the smallest of the three there is only about one chance in one thousand that a difference this great will be obtained by chance sampling errors when the true difference is zero. The fourth ratio is too small to be considered significant. It should be noted that the ratio for type D errors was too small on all the tests to be considered definitely significant.

The 65-student group appeared to be somewhat superior to the 49-student group on the three background subjects measured, but the 49-student group was definitely superior to the 65-student group on the first three items of Test A (Errors A, B, and C). In view of this, it seems reasonable to assume that the superior performance of the 49-student group was due largely to the methods by which they were taught.

# CHAPTER IX

## SUPPLEMENTARY TESTS

As PREVIOUSLY stated it is extremely difficult to classify the types of errors in complex problems, such as combined multiplication and division. Test C was given in order to compare the number of incorrect answers in simple examples with the number of incorrect answers in more complex examples. In Test C there are nine examples in simple multiplication and division (Part I) and nine in combined multiplication and division (Part II).

### TEST C

#### Part I

1. $5.6 \times 0.359 =$
2. $10.63 \times 8.81 =$
3. $36.7 \times 0.084 =$
4. $144 \times 7.26 =$
5. $231 \times 47.5 =$
6. $0.8637 \times 4.47 =$
7. $0.00062 \times 9440 =$
8. $3.14 \times 1.86 =$
9. $0.927 \times 0.0043 =$

#### Part II

1. $\frac{54.73 \times 6.02}{8.77} =$
2. $\frac{3173 \times 0.622}{0.000037} =$
3. $\frac{3.5 \times 1728}{231} =$
4. $5.97 \times 0.0046 \times 92.3 =$
5. $\frac{14.3 \times 0.0065}{27.6} =$
6. $\frac{3.6 \times 231}{1728} =$
7. $\frac{1}{2} \times 36.8 \times 12.5 =$
8. $\frac{4.43}{6.97 \times 4.08} =$
9. $\frac{0.046 \times 33.8 \times 9.26}{1553 \times 70.6} =$

Test C was given to 119 graduate students at Teachers College, Columbia University. These students did not take Test A. The results of this test were as follows.

Number of incorrect answers in Part I ....... 174
Number of incorrect answers in Part II ...... 170

These results indicate that the students made about the same number of errors in complex examples as they did in simple examples. The results of Test C would seem to justify the use of the simple examples of Test A as a basis for classification of errors.

## SUPPLEMENTARY TESTS D, E, F, AND G

A series of four other tests were given in order to make a further study of certain errors and to determine the relative difficulty of different cases of the same operation. These tests were given to approximately 280 high school juniors and seniors and college freshmen by friends of the author. All these students took tests D, E, F, and G. The tests had examples in square root, multiplication, and division. In these latter tests an attempt was made to answer the following questions:

1. What is the relative number of errors in reading the various types of graduations?
2. Are more errors made in finding square roots when the graduations of the given number and its square root are the same, or when they are different?
3. Are more errors made when the second significant figure of an answer is zero than when the second figure is some other digit?
4. Are more errors made in multiplication when the product is found to the left of the first factor than when the product is found to the right of the first factor?
5. Are multiplication and division more likely to be confused in problems where simple inspection will not clearly distinguish between the product and the quotient?
6. How does slide rule computation compare with pencil computation?

The problems in the tests were generally restricted to finding the square root of a given number, or the quotient or product of two numbers. In such simple problems it is relatively easy to determine the specific causes of incorrect answers. For example, if the value of $\sqrt{0.00455}$ is given as 0.0214, it is obvious that the student has found the square root of $\sqrt{0.0455}$. If the quotient of 435 ÷ 112 is given as 4.88, the student has confused multiplication and division. In more complex problems, such as $\frac{34.2 \times 117.5}{398 \times 577}$, many operations and settings are involved, and it becomes very difficult to determine the exact cause of an incorrect result.

## TEST D, PART I: ERRORS IN ESTIMATING THE THIRD SIGNIFICANT FIGURE ON THE D SCALE

Test D consisted of eighteen examples in square root. Since the purpose of the first six examples was different from that of the later twelve, this test will be discussed in two parts.

### TEST D: PART I

Find the square roots of the following numbers:

| | Number | Answer | | Number | Answer |
|---|---|---|---|---|---|
| 1. | 19 | 4.36 | 4. | 35 | 5.92 |
| 2. | 4.5 | 2.12 | 5. | 1.9 | 1.38 |
| 3. | 2.5 | 1.58 | 6. | 13 | 3.61 |

The given numbers were so selected that two answers were between the primary graduations 1 and 2 on the D scale, two answers were between 2 and 4, and two answers were between 4 and 10. The purpose of Test D, Part I, was to determine the effect of the three types of graduations of the D scale on the estimation of the third significant figure of the answer. Given numbers having only two significant figures were selected in order to reduce the chances of error in setting the given number. Preliminary drill was given on distinguishing $\sqrt{4.5}$ and $\sqrt{45}$, etc. Table XVIII shows the number of errors made by 280 students.

Table XVIII also shows that the number of errors increased as the number of tertiary D scale graduations became smaller. It is of interest to note the relatively large number of errors between the primary graduations 2 and 4. Between the primary graduations 4 and 10, it is not always possible to obtain the precise value of the third significant figure. Between 2 and 4, however, the third significant figure is either located by a tertiary graduation or found by bisecting the space between two such graduations. The results of this test indicate, therefore, that more attention should be given to answers occurring between 2 and 4.

TABLE XVIII

Results of Test D, Part I: Errors in Estimating the Third Significant Figure

| Answer between Primary Graduations | Number of Errors in Third Significant Figure |
|---|---|
| 1 and 2 | 27 |
| 2 and 4 | 125 |
| 4 and 10 | 198 |

## TEST D, PART II: ERRORS CAUSED BY CONFUSION OF GRADUATION ON THE A AND D SCALES

It is to be noted that the graduations in corresponding parts of the A and D scales are generally different. For example, the square root of any number between the primary 1 and 2 of the A scale will fall between 1 and 1414 of the D scale. In this case the tertiary graduations on the A and D scales are different, since they occur at intervals of 2 on the A scale, and at intervals of 1 on the D scale. There are two instances where both a number and its square root will have the same type of tertiary graduation, namely, if the given number is between the primary 10 and 16 or between the primary 20 and 50.

The twelve examples of Test D, Part II, were designed to determine whether more errors are made in square root when the tertiary graduations of the A and D scales are different than when the graduations are of the same type.

## TEST D: PART II

Find the square roots of the following numbers:

| | Number | Answer | | Number | Answer |
|---|---|---|---|---|---|
| 7. | 1.64 | 1.28 | 13. | 14.6 | 3.82 |
| 8. | 3.75 | 1.936 | 14. | 46.5 | 6.82 |
| 9. | 14.2 | 3.77 | 15. | 18.2 | 4.27 |
| 10. | 21.5 | 4.64 | 16. | 2.35 | 1.53 |
| 11. | 4.45 | 2.11 | 17. | 28.5 | 5.34 |
| 12. | 19.4 | 4.40 | 18. | 11.6 | 3.42 |

The twelve problems are "paired," 7 with 13, 8 with 14, etc. For example, the given numbers in 7 and 13, 1.64 and 14.6, are found on the same type of graduation on the A scale, and both numbers fall exactly on the tertiary graduation. In 7 the answer, 1.280, is found on a different type of graduation from 1.64; in 13, answer and given number are both found on the same type of graduation.

The results given in Table XIX show that when the student was required to read two different types of graduations in the same problem, the number of errors was about three times as great as the number of errors when the graduations were similar.

TABLE XIX

Results of Test D, Part II: Confusing Graduations on A and D Scales

| Graduations on A and D Scales | Number of Errors in Estimating Third Significant Figure |
|---|---|
| Same | 22 |
| Different | 75 |

The examples in Test D were intentionally made so simple that the first significant figure of the root could readily be determined by inspection.

## TEST E: ZERO ERRORS

Test E was given in order to make a further study of zero errors. In this test there are six examples in division of two-place numbers.

Three examples have the second significant figure of the answer zero, and three do not. Two-place numbers were selected in order to reduce errors in setting the given numbers. The answers are made up so that each answer falls exactly on a tertiary graduation, in order to avoid estimating the third figure. Problems are paired, 1 with 4, 2 with 5, and 3 with 6. In problems 1 and 4, 77 is the same type of graduation as 66, 37 the same graduation as 28, both answers are read under the left index, and both answers fall on the same type of graduation. The decimal points in all answers may be readily located by inspection. Test E was given to 270 students.

### TEST E

Find the following quotients:

| | Example | Answer | | Example | Answer |
|---|---|---|---|---|---|
| 1. | $\frac{77}{37}$ | 2.08 | 4. | $\frac{66}{28}$ | 2.36 |
| 2. | $\frac{67}{54}$ | 1.24 | 5. | $\frac{53}{51}$ | 1.04 |
| 3. | $\frac{85}{21}$ | 4.05 | 6. | $\frac{74}{17}$ | 4.35 |

The results of Test E confirmed in a rather striking manner the results of Test A (given to the groups of 660 and 150 students) with respect to zero errors. There were 171 zero errors in the problems having zero as the second significant figure of the answer, and only 27 where the second figure was not zero. The results of Test E re-emphasized the fact that special attention must be given to zero errors.

### TEST F: "RIGHT" VERSUS "LEFT" MULTIPLICATION

There are two cases of slide rule multiplication as follows: (1) the right index of the C scale may be placed in line with the first factor and the answer read to the left; or (2) the left index may be used and an answer found to the right. Test F was given to determine whether more errors are made in multiplication when the right index is used than when the left index is used.

In multiplication, if both of the given numbers fall between the primary graduations 1 and 2, the left index is always used, since the answer must fall between the primary 2 and 4. Since either factor may fall on any one of three types of graduations, there are six cases as follows:

| Case | First Factor Between Primary | Second Factor Between Primary | Index Used |
|---|---|---|---|
| 1. | 1 and 2 | 1 and 2 | left |
| 2. | 1 and 2 | 2 and 4 | left |
| 3. | 1 and 2 | 4 and 10 | right or left |
| 4. | 2 and 4 | 2 and 4 | right or left |
| 5. | 2 and 4 | 4 and 10 | right or left |
| 6. | 4 and 10 | 4 and 10 | right |

In Test F there are twelve examples, six in which the right index is used and six in which the left is used. The examples include cases 3, 4, and 5. In examples 1 and 7, the given numbers fall on the same type of graduation; in example 1 the left index is used, and in example 7 the right index. Similarly, example 2 is paired with example 8, 3 with 9, etc. Test F was given to 276 students.

## TEST F

Find the following products:

1. $3.64 \times 2.28 = 8.30$
2. $72.5 \times 1.22 = 88.4$
3. $21.6 \times 42.5 = 918.$
4. $18.3 \times 51.5 = 9420$
5. $28.6 \times 25.8 = 732$
6. $74.5 \times .172 = 12.81$
7. $31.6 \times 36.4 = 1150$
8. $.785 \times 1.86 = 1.461$
9. $3.44 \times 5.35 = 18.40$
10. $174 \times 7.65 = 1331$
11. $3.94 \times 3.38 = 13.32$
12. $47.5 \times 16.4 = 770$

It was expected that there would be more errors when the right index was used, so the results of the test were rather surprising. There were 465 errors when the left index was used, and 427 when the right index was used. This difference is probably not significant, and the only conclusion warranted is that it does not seem to increase the difficulty when the right index is used.

### TEST G: CONFUSION OF MULTIPLICATION AND DIVISION

The operations of multiplication and division are sometimes confused, and the division answer is given to an example in multiplication and vice versa. Generally simple inspection will distinguish between the results of multiplication and division. The product $ab$ and the quotient $a \div b$ tend to have the same significant figures as $b$ approaches $(10)^n$, or $\sqrt{10}$. For example,

(1) $525 \div 98.6 = 5.32$, and $525 \times 9.86 = 5176$.
(2) $525 \div 31.8 = 16.5$, and $5.25 \times 3.18 = 16.7$.

Test G is intended to obtain a measure of the confusion of division and multiplication, especially where inspection does not clearly differentiate the quotient from the product. Test G was given to 272 students.

### TEST G

Find the answers to the following examples in multiplication and division:

| | Example | Answer | | Example | Answer |
|---|---|---|---|---|---|
| 1. | $\frac{23.2}{3.24}$ | 7.16 | 7. | $\frac{21.4}{3.86}$ | 5.55 |
| 2. | $23.6 \times 1.12$ | 26.4 | 8. | $37.8 \times 1.54$ | 58.2 |
| 3. | $2.56 \times 9.25$ | 23.7 | 9. | $2.32 \times 6.25$ | 14.5 |
| 4. | $\frac{34.8}{112}$ | 0.308 | 10. | $\frac{38.4}{17.6}$ | 2.18 |
| 5. | $\frac{18.3}{61.5}$ | 0.298 | 11. | $\frac{16.6}{91.5}$ | 0.1814 |
| 6. | $0.675 \times 32.3$ | 21.7 | 12. | $0.555 \times 21.2$ | 11.75 |

These twelve examples were paired after the manner of Tests E and F. In examples 2 and 8, 23.6 and 37.8 fall on the same type of graduation as do 1.12 and 1.54. The answers to 2 and 8 are both found by using the left index. In example 2, by inspection, $23.6 \times 1.12 = 2–.–$, and $23.6 \div 1.12 = 2–.–$. In example 8, however, by inspection, $37.8 \times 1.54 = 6–.–$, but $37.8 \div 1.54 = 2–.–$. Similarly, example 1 is paired with example 7, 3 with 9, etc. In problems

1, 2, 3, 4, 6, and 11 the inspection answers for multiplication and division may be confused; in other problems the approximations made by inspection are distinct.

TABLE XX

Results of Test G: Confusing Multiplication and Division

| Error | Problems 1, 2, 3, 4, 6, 11 | Problems 5, 7, 8, 9, 10, 12 |
|---|---|---|
| Division for multiplication......... | 13 | 7 |
| Multiplication for division......... | 70 | 4 |
| Total...................... | 83 | 11 |

Since there were only 11 errors in the "obvious" problems and 83 in the others, it is safe to conclude that multiplication and division are likely to be confused by beginning students when the answer cannot be readily checked by inspection. Confusion of this type may be reduced by drill on multiplication and division of one- and two-digit numbers.

In 27 cases in Test G, the student found the dividend, placed the index over it as if to multiply, and then read the answer under the index, that is, $16.6 \div 91.5 = 16.6$.

In order to compare the efficiency of slide rule computation and pencil computation, 59 of the students who had taken Test G were asked to repeat this same test on paper. The results of the pencil test seem to offer a strong argument for the use of the slide rule. The 59 students had made 74 errors when they used the slide rule and they made 139 errors in pencil computation.

## SUMMARY OF CHAPTER IX

The author's procedure of restricting Test A to simple examples seems to have been justified by the results of Test C. On this test the students made about the same number of errors on the simple examples as they made on the complex examples.

In Tests D to G it was found that: (1) The number of errors was greater in the part of the scales where there were fewer tertiary graduations. (2) The errors were increased when more than one type of tertiary graduation occurred in the same problem. (3) There were more errors when the second significant figure of the answer was zero. (4) Beginning students confused multiplication and division in problems where inspection did not distinguish between quotient and product.

# CHAPTER X

## GENERAL SUMMARY

THIS study shows that the answers obtained by solving problems on the slide rule are as accurate as most of the measurements made in industry and trade, and that slide rule solutions are satisfactory for a very large per cent of the problems met in fairly technical scientific work.

The study has shown that students must have certain prerequisites for the successful learning of the slide rule. These are (a) a thorough working knowledge of decimal fractions, (b) habitual use of the rules for approximate computation, and (c) an appreciation of the value of answers correct to two, three, or four significant figures.

The study has also shown how recent improvements in the teaching of decimal fractions, the increase in the use of the metric system, the wide tendency to divide many of our old measuring units decimally, and the recognition of the importance of approximate computation have cleared the way for a wider use of the slide rule. Fortunately, satisfactory slide rules are now produced in large quantities at prices that offer no serious financial handicap to their more extended use.

An analysis of the answers to more than 27,000 examples worked on the slide rule show that four classes of errors included a very large percentage of all common slide rule errors (about 93%). These common errors are: (a) errors in locating the decimal point in answers, (b) errors in estimating the third significant figure of the answer, (c) zero errors, and (d) errors in reading the tertiary slide rule graduations. Ten other miscellaneous errors account for the remaining 7% of the total errors.

Special teaching techniques greatly reduced the number of errors of the first three types, and seemed to help reduce the number of errors of the fourth type when these techniques were used in teaching an experimental group of 150 students. These techniques are: (1) the use of the standard number method for locating the decimal point where the decimal point cannot be located at once by simple inspection, (2) starting slide rule computation with two-figure data and answers and then giving drill with three- and four-figure data after the students have gained some mastery of the slide rule, (3) giving special drill with data and answers having zero as a second figure in order to make and keep the student "zero conscious," and (4) drill in estimating the third or fourth significant figures of the answer.

The students tested made about as many errors in simple exercises as they did in more complex work. This was partly due to the fact that they developed the habit of using the standard number method of locating the decimal point in complex work and of depending on inspection in the simple exercises. The students who had learned to operate the slide rule efficiently made a smaller percentage of errors in slide rule work than they did in pencil computation, and the errors in slide rule work were far less serious than those made in pencil computation. The students very seldom had errors in the first two figures of their slide rule answers. In pencil computation the first figures of an answer were as likely to be incorrect as the last figures.

## EDUCATIONAL IMPLICATIONS

In concluding this study, the following recommendations are made:

1. That the standard number method be used for locating the decimal point in all slide rule instruction.

2. That considerable drill be given early in slide rule instruction with examples having zero as a second or third significant figure in the examples or the answers.

3. That the early instruction on the slide rule be limited to one and two significant figures in both data and answers.

4. That considerable drill be given in reading the third significant figure in slide rule answers, as suggested on pages 60 to 63.

5. That the slide rule should not be introduced before the middle

of the ninth year, as it is not possible to teach efficiently the standard number method before this time.

6. That all textbooks and tests should show by the data used in the problems the number of significant figures desired in the answers and that the answers in all specimen solutions and answer books should be consistent with the data given in the problems.

7. That the fundamental concepts of approximate computation be taught before any attempt is made to teach the slide rule.

As an extension of this study, further research should determine the following:

1. What is the most satisfactory grade placement for elementary slide rule instruction?

2. Will students who have been taught to use the slide rule in the ninth grade, and who have used it during the tenth, eleventh, and twelfth grades be more proficient in mathematical subjects than those who have not used the slide rule?

3. Can the elementary concepts and principles of approximate computation be taught in grades six to nine?

4. What is the most satisfactory grade placement of the various topics of approximate computation?

5. What professionalized subject matter is necessary for efficient teacher training in slide rule instruction?

6. Will students who have used the slide rule for the solution of the various problems in physics and chemistry be more efficient in these subjects than students who have not used the slide rule?

# BIBLIOGRAPHY

## I. General References


Adams, Daniel. *Scholars Arithmetic.* John Prentis, Keene, N. H., 1807. p. 164.

Bakst, A. *Approximate Computation.* Twelfth Yearbook, National Council of Teachers of Mathematics. Bureau of Publications, Teachers College, Columbia University, New York, 1937. pp. 287.

Ball, W. W. R. *A Short Account of the History of Mathematics.* Macmillan and Co., London, 1888.

Brooks, Edward. *Normal Standard.* Christopher Sower Co., Philadelphia, 1895.

Cajori, F. A. *A History of Elementary Mathematics.* Macmillan and Co., New York, 1896.

Cajori, F. A. *A History of the Logarithmic Slide Rule and Allied Instruments.* Engineering Publishing Co., New York, 1909.

Davies, Charles. *University Arithmetic.* A. S. Barnes and Co., New York, 1866.

Dilworth, Thomas. *Schoolmasters Assistant.* R. Aitken, Philadelphia, 1781.

Fink, E. *A Brief History of Mathematics.* Open Court Publishing Co., Chicago, 1890.

Greenleaf, Benjamin. *National Arithmetic.* Sower Barnes and Co., Philadelphia, 1860.

Greenleaf, Benjamin. *New Practical Arithmetic.* Robert S. Davis Co., Boston, 1873.

Harding, T. Swann. "What Is Scientific Exactitude?" *Atlantic Monthly,* April, 1931.

Hodgman, Charles D. and Lange, Norbert A. *Handbook of Physics and Chemistry.* Chemical Rubber Publishing Co., Cleveland, Ohio, 1936.

Lundell, G. E. "The Chemical Analysis of Things as They Are." *Industrial and Engineering Chemistry* (Analytical Edition), Vol. 5, No. 4, July 15, 1933.

Roller, Duane. *Terminology of Physical Science.* University of Oklahoma Press, 1936. p. 36.

Runge, C. and Koenig, H. *Vorlesungen über Numerisches Rechnen.* Julius Springer, Berlin, 1924.

Sanford, Vera. *Short History of Mathematics.* Houghton Mifflin Co., Boston, 1930.

Smith, D. E. *History of Mathematics,* Vols. I and II. Ginn and Co., Boston, 1923.

Snedecor, George. *Statistical Methods.* Collegiate Press, Inc., Ames, Iowa, 1938.

Strayer, George D. and Upton, C. B. *Arithmetic for the Higher Grades.* American Book Co., New York, 1928. p. 248.

Thompson, J. B. *New Practical Arithmetic.* S. C. Griggs and Co., New York, 1876.

Thompson, J. E. *A Textbook in Algebra.* D. Van Nostrand Co., Inc., 1935.

Upton, C. B. "Changing the Curriculum in Arithmetic." *Teachers College Record,* December, 1926, pp. 46-48.

Vodges, Wm. *United States Arithmetic.* Caleb Bartlett, Philadelphia, 1845.

Wentworth, G. A. *Advanced Arithmetic.* Ginn and Co., Boston, 1898. pp. 29-55.

Yule, G. Undy and Kendall, M. G. *An Introduction to the Theory of Statistics.* Chas. Griffin and Co., Ltd., London, 1937.

## II. Slide Rule Manuals

Black, R. D. and M'Cord, Mary. *Introducing the Slide Rule.* Nixon Enterprises, Wabash, Indiana, 1938.

Breckenridge, W. R. *The Polyphase Slide Rule.* 1925. *The Log Duplex Slide Rule.* 1936. *The Mannheim Slide Rule.* 1925. Keuffel and Esser Co., New York.

Breslich, E. R. and Stone, C. A. *The Slide Rule.* University of Chicago Press, Chicago, 1929.

Cooper, H. O. *Slide Rule Calculations.* Oxford University Press, London, 1931.

Cox, William. *The Mannheim Slide Rule.* Keuffel and Esser Co., New York, 1891.

Cullimore, A. R. *The Phillips Slide Rule.* Eugene Dietzgen and Co., New York, 1925.

Dietzgen, Eugene and Co. *The Maniphase Slide Rule.* Eugene Dietzgen and Co., New York, 1938.

Gilson Slide Rule Co. *The Midget 5-in-1 Slide Rule.* Gilson Slide Rule Co., Stuart, Fla., 1925.

Keal, H. M., Phelps, W. C., and Leonard, C. J. *The Slide Rule.* John Wiley and Sons, New York, 1932.

Keuffel and Esser Co. *Elementary Instructions for Operating the Slide Rule.* Keuffel and Esser Co., New York, 1938.

Nestler, A. *Nestler's Duo Multiplic and Duo Log Slide Rules.* Albert Nestler, Baden (date not given).

Richardson, Geo. W. and Clark, J. J. *The Slide Rule Simplified.* Gilson Slide Rule Co., Stuart, Fla., 1918.

Rohrberg, A. Theorie und Praxis des Logarithmuschen Rechenstabes. V. B. G. Teubner, Leipzig, 1925.

Rosenthal, L. W. *Mannheim and Multiplex Slide Rules.* Eugene Dietzgen and Co., New York, 1911.

Sainte-Lague, A. M. *La Regule a Calcul.* Paris, 1934, pp. 34.

Shireby, R. M. *The Slide Rule Applied to Commercial Calculations.* Sir Isaac Pitman and Sons, Ltd., New York, 1930.

Thompson, J. M. *A Manual of the Slide Rule.* D. Van Nostrand Co., Inc., New York, 1930.

III. UNITED STATES GOVERNMENT DEPARTMENT OF COMMERCE PUBLICATIONS

1. Federal and State Laws Relating to Weights and Measures (No. 20), Third Edition, 1926. pp. 1-976.
2. Specifications and Tolerances for Commercial Weighing and Measuring Devices (No. 85), 1929. pp. 1-130.
3. Method for Precision Test of Large Capacity Scales (No. 199), 1921. pp. 3-16.
4. History of the Standard Weights and Measures of the United States (No. 64), 1925. pp. 111-134.
5. Legal Weights in Pounds per Bushel of Various Commodities (No. 10), Fourth Edition, 1924. pp. 1-18.
6. Weighing by Substitution (No. 208), 1922. pp. 177-192.
7. Specifications for Marine Sextants (No. 110), 1921. pp. 3-8.
8. Buying Commodities by Weight or Measure (No. 45). pp. 3-42.
9. Testing of Line Standards of Length (No. 332), 1927. pp. 1-22.
10. A Fundamental Basis for Measures of Length (No. 535), 1926. pp. 395-408.
11. Testing of Measuring Tape (No. 328), 1927. pp. 1-16.
12. Testing Precision of Gage Blocks (No. 436), 1922.
13. The International Metric System of Weights and Measures (No. 2), 1922. pp. 2-16.
14. Weights and Measures Administration, 1927. pp. 1-270.
15. Michelson, A. A. *Measurement of the Velocity of Light between Mt. Wilson and Mt. San Antonio.* Contributions from the Mt. Wilson Observatory (No. 329), 1927. pp. 1-22.

# APPENDIX

## INTRODUCTION TO THE TEACHING OF THE SLIDE RULE

As pointed out in the preceding chapters, students should be taught thorough respect for two-, three-, and four-figure accuracy, should be able to use standard numbers, and should understand the techniques of approximate computation before any attempt is made to teach the slide rule. The above concepts and techniques can be taught early in the high school and are important enough to be taught even when no attempt is made to teach the slide rule.

In suggesting methods for teaching the slide rule, it is assumed that the students will be well grounded in the topics listed above. It is also assumed that instruction in slide rule computation will be begun early in high school mathematics and that this instruction will be continued in the various other high school courses in mathematics. It is further assumed that the students will be given plenty of opportunity for slide rule computation, and that in some of the work in mathematics and science the slide rule solution will be the only solution required.

In teaching any new thing it is very important that the first lessons be so interesting and valuable that a "sale" is at once made. There are a number of different ways in which the slide rule may be presented. In the seventeen different slide rule manuals listed on page 97, nine begin with the reading of primary, secondary, and tertiary divisions of the scales. In four other manuals the slide rule is introduced as an application of logarithms. Two begin with a history of the slide rule and two others introduce the slide rule with simple one-figure multiplication and division, such as $3 \times 2 = 6$ and $8 \div 2 = 4$. In these last two manuals the next step is the extension of multiplication and division to three-figure numbers.

The author asked 103 teachers how they began the teaching of the slide rule. Eighteen of these teachers began with logarithms, 30 with reading the scales, 53 with simple multiplication and division, and 2 with the history of the slide rule.

The author has tried all of these methods and has discarded them for the reasons given below.

If the slide rule is introduced through logarithms, the teaching of the slide rule must be delayed until the third or fourth year of the high school, or until after logarithms have been taught in intermediate algebra or trigonometry. This delay in the teaching of so valuable a mathematical instrument is unfortunate. The slide rule is one of the best means we have for "selling mathematics" to high school pupils and should be taught as early in high school as possible. Also, there is no reason why a pupil should be forced to wait until the theory back of the construction of the slide rule can be given. Very few people who use the various computing instruments employed in commercial computation know the theory back of their construction, or know that they are simply mechanical abaci. Few people who drive automobiles, or use electric refrigerators and other inventions so essential to modern life, know the science back of their construction. There is little evidence that a knowledge of the theory back of the slide rule will help a student to become a more efficient operator. It would seem better to begin the slide rule early in the high school and then, when the student is learning logarithms in the eleventh or twelfth grade, use the slide rule to motivate logarithms. In fact, it is very easy to teach logarithms to a class if the class has previously learned to use the slide rule, and the standard number method for locating the decimal point. Also, if the teaching of the slide rule is delayed until after the study of logarithms, many students who never get as far as this in mathematics will be denied the opportunity to learn to use the slide rule.

To begin slide rule instruction with the reading of all the various scales to three figures seems undesirable. Several days may be spent on work that is not at all interesting and the best first impressions are thereby sacrificed.

There are also objections to starting the work with the slide rule with simple multiplication and division. A high school student is not greatly impressed with the fact that $3 \times 2 = 6$.

Even two- and three-figure multiplication and division may seem like primary work to a high school student. A student tries to find the product of 236 and 384 on the slide rule. After a long time he reads on the rule 905, but thinks the answer may be 9050. As a check he uses a pencil, and in much less time than he required for the slide rule solution, he finds the answer to be 90624. Such an experience will not give him a very good opinion of the slide rule as an aid in computation.

There are two methods of introducing the slide rule that seem better than any of the methods given above. The first of these ways is through the teaching of square root.

Square root is an interesting new operation. A paper and pencil solution is difficult, while a slide rule solution is very easy. It is not necessary to manipulate the slide, and square root is therefore not as complex an operation on the rule as multiplication or division. Square root also furnishes a splendid opportunity to teach students to read the slide rule scales and to locate or set numbers on the rule. Also, even with the pencil solution, the answers to practically all square root problems are approximate. Because of this, students are not disturbed by the approximate solutions obtained on the slide rule.

A second method of introducing the slide rule is through the teaching of proportion. If the proportion,

$$\frac{x}{6} = \frac{4}{8} \qquad \frac{x(\mathrm{C})}{6(\mathrm{D})} = \frac{4(\mathrm{C})}{8(\mathrm{D})}$$

is set up on the C-D scales of the slide rule exactly as it is written above, 8 on the D scale with 4 directly over it on the C scale, the answer 3 will be found on the C scale directly over 6 on the D scale, as shown by the second proportion above.

Proportions may be written in the following manner:

$$\frac{x}{6} = \frac{4}{8}; \ \frac{3}{x} = \frac{4}{8}; \ \frac{3}{6} = \frac{x}{8}; \text{ and } \frac{3}{6} = \frac{4}{x}.$$

Changing the position of $x$ does not increase the difficulty of setting up on a slide rule. At the start all proportions should be of the kind that will not require the shifting of an index, as would be necessary in the following:

$$\frac{x}{3} = \frac{2}{8} \text{ or } \frac{x}{6} = \frac{7}{3}.$$

If the slide rule is introduced through proportion, it is advisable to begin with one- or two-figure data and two-figure answers.

Many slide rule manuals suggest that proportion be solved in the following manner.

Example: $$\frac{x}{3.42} = \frac{19.1}{61.3}$$

Solution (A): Multiplying both sides by 3.42 you have

$$x = \frac{19.1 \times 3.42}{61.3}$$

Solve by multiplication and division.

Instead of using this method, it is much better to "write" the proportion on the rule exactly as it appears on paper and read the answer at once without any calculation. The solution would then be as follows:

Solution (B): $$\frac{19.1}{61.3} = \frac{x}{3.42}$$

In this solution 19.1 is first set over 61.3 and the answer is read directly over 3.42.

The method just described is also much more efficient in problems involving continued proportion, as is shown by the paired solutions below. Solution B is the preferable solution.

Example I: Given the triangle $ABC$ with $\angle A = 70°$, $\angle B = 60°$, and side $c = 375$ ft. Find sides $a$ and $b$.

Solution (A):

$$\frac{375}{\sin 50} = \frac{a}{\sin 70} \qquad \frac{375}{\sin 50} = \frac{b}{\sin 60}$$

$$a = \frac{375 \times \sin 70}{\sin 50} \qquad b = \frac{375 \times \sin 60}{\sin 50}$$

Solution (B).

$$\frac{375}{\sin 50} = \frac{a}{\sin 70} = \frac{b}{\sin 60}$$

In solution (B), the answers $a$ and $b$ are found at once over 70° and 60° with much less effort than in solution (A).

In many business problems it is necessary to divide a sum of money into amounts proportional to a given set of numbers. This is similar to the topic of partnership and is called prorating. When a cargo is sent over several railroads, the total freight bill is prorated according to the mileage the cargo traveled on each system.

Example II: Prorate $125.00 in proportion to the following mileages: 630 mi., 420 mi., 280 mi., 160 mi. (total mileage, 1490 mi.).

Solution (A):

$$\frac{x}{630} = \frac{125}{1490} \text{ or } x = \frac{630 \times 125}{1490}$$

$$\frac{y}{420} = \frac{125}{1490} \text{ or } y = \frac{420 \times 125}{1490}$$

$$\frac{z}{280} = \frac{125}{1490} \text{ or } z = \frac{280 \times 125}{1490}$$

$$\frac{w}{160} = \frac{125}{1490} \text{ or } w = \frac{160 \times 125}{1490}$$

Solution (B):

$$\frac{125}{1490} = \frac{x}{630} = \frac{y}{420} = \frac{z}{280} = \frac{w}{160}$$

The answers $x$, $y$, $z$, and $w$ may be read at once from the rule.

It often happens that three, four, five, or more terms are required and all can be read almost instantly from the slide rule.

If the instructor wishes to show the efficiency of the slide rule as a computing instrument, no topic is better than proportion. For an easy, interesting, and valuable introduction, proportion is unique.

Square root and proportion are interesting topics. They are more difficult than simple multiplication and division. They are identified with high school mathematics rather than with the mathematics of the elementary school. If, after a few minutes' instruction, a student finds he can extract square root with his slide rule, or solve a proportion (equation) on it, he is almost certain to form a favorable opinion of the slide rule as an aid to computation, and be in the proper frame of mind to do the necessary drill work.

# VITA

CARL N. SHUSTER was born in Frenchtown, N.J., Feb. 16, 1890. He received his elementary education in the public schools of Everittstown, N.J., and his secondary education at the Frenchtown, N.J. high school and the Model High School, Trenton, N.J. He attended the State Normal School at Trenton, N.J. during 1910, 1911, and 1912, and Teachers College, Columbia University during 1913 and 1914. From the latter institution he received the degree of Bachelor of Science in 1914 and that of Master of Arts in 1918. He continued his graduate study at Teachers College from 1921 to 1929.

He taught from 1907 to 1910 in the public school of Wincel's Grove, N.J. Since that time he has held the following positions: teacher, high school, Dover, N.J., Feb. to June, 1913; teacher, high school, Westfield, N.J., 1914; head of the mathematics department, Clark School, New York, 1915 to 1917; United States Navy, 1917 to 1918; senior cost accountant, United States Civil Service, 1918 to 1920; teacher of mathematics, Bowling Green University, Bowling Green, Ohio, summer sessions of 1920 and 1921; head of the mathematics department, high school, Orange, N.J., 1920 to 1929; head of the continuation school for boys, Orange, N.J., 1925 to 1929; teacher, Pennsylvania State College, State College, Pa., summer session 1926; instructor in mathematics, Teachers College, Columbia University, 1926– ; professor of mathematics and department head, New Jersey State Teachers College, Trenton, N.J., 1929–

During the years 1907 to 1910, he worked out a full project curriculum with integrated subject matter for rural schools; he gave a number of talks on this in various Teachers College classes in 1913 and 1914. In 1935, he and Mr. Fred Bedford published the book "Field Work in Mathematics" (American Book Co.). In 1938, he collaborated with A. E. Bond, R. G. Reynolds, and R. L. West in writing the Real Life series of arithmetics (Charles Scribner's Sons).

# Power, Passion, and Prejudice

## Shootout in the Galiuro Mountains

Barbara Brooks Wolfe

IMAGO PRESS
TUCSON ARIZONA

Published in the United States of America by:

Imago Press
3710 East Edison
Tucson AZ 85716

Library of Congress Cataloging-in-Publication Data

Wolfe, Barbara Brooks, 1928-
Power, passion, and prejudice : shootout in the Galiuro Mountains / Barbara Brooks Wolfe.
p. cm.
Includes bibliographical references and index.
ISBN-13: 978-0-9799341-7-9 (pbk. : alk. paper)
ISBN-10: 0-9799341-7-6 (pbk. : alk. paper)
1. Criminal justice, Administration of—Arizona—Greenlee County. 2. Trials (Murder)—Arizona—-Greenlee County. 3. Violence—Arizona—Galiuro Mountains. 4. Galiuro Mountains (Ariz.) [proposed]—History. I. Title.
HV9955.A7W65 2008
364.152'3092--dc22

2008040704

Book and Cover Design by Leila Joiner

Cover photograph "antique map, Tucson Arizona" [circa 1916] © Robert Kyllo
Cover photograph "prison" © danimages

ISBN 978-0-9799341-7-9
ISBN 0-9799341-7-6

Printed in the United States of America on Acid-Free Paper

This book is dedicated to my friend and mentor,
Margaret Scariano,
who inspired me to write this story.

# Acknowledgements

Without the help of Dr. Richard Froede and the late Dr. Alan Jones, this book could not have been written. For their scientific curiosity and their intellectual generosity, I am forever grateful.

I also want to acknowledge the contributions of W. R. (Ryder) Ridgeway. Anyone writing Graham County history owes a debt of gratitude to Ryder, and I am no exception. Ryder's interviews with old-timers, his long habit of collecting and preserving photos, documents, and written records are invaluable to historians and writers. He generously shared these treasures with any interested party, and I am eternally grateful for his work.

Russell Murdock, my old friend and classmate, had first-hand knowledge of the Aravaipa region and many of the people who lived there. Memories he shared with me were essential to this book.

I also want to thank Klondyke rancher Ray Claridge, the Haby twins, and the women in the Greenlee County Courthouse who were so helpful to me in locating records.

For others who made valuable contributions to this book and prefer to remain anonymous, I offer my heart-felt thanks.

In times of calamity, any rumor is believed.

—Publilius Syrus: *Maxims*

# Preface

In 1918 a legendary shootout in Arizona made headlines in newspapers across the nation. According to the press, German sympathizers had murdered a federal posse in the Galiuro Mountains in Graham County. Misinformation on the case reported in print would fill volumes, but four men died in that gun battle, and the bitter anger that followed reintroduced the death penalty to Arizona and divided the state for nearly a century.

Born and raised in Graham County, I was familiar with the legend and knew most of the families involved. Decades have passed since the shootout, but the anger of surviving families carried down to new generations and did not cool with time. How could it be, I wondered, that this case left so many unanswered questions and so much bitterness?

In 1985 I began to research the case. It seemed to me the time had come to set the legend aside and pay a debt of truth to history. Too many lives had been ruined by this tragedy. It was time to examine facts, face the truth, and learn what we could from it. I had no idea how difficult that task would be.

Interviews with old-timers produced conflicting information. I had better luck with medical and legal experts, but I needed documentation. In Greenlee County, where the alleged German sympathizers were tried for murder, the Court Clerk informed me there was no trial transcript; the court reporter's notes had disappeared. Unwilling to accept that fact, I searched through tons of paper in old files housed in the abandoned jail in Greenlee County.

It was August. This part of the courthouse had no air conditioning, and the windows were nailed shut. Women who worked in the building offered me an electric fan, but no outlets were available in the old jail. I had to work in five and ten minute cycles—ten minutes to search and five minutes in the hall to cool off. This arduous procedure produced nothing but heat exhaustion and disappointment.

Elsie Sims, the Court Clerk, discovered a vault outside the courthouse that contained old trial records. No one had the combination to the locked metal door, which had survived numerous floods and was rusted shut. After Elsie called a welder to open the vault, I eagerly searched through hundreds of legal papers stained with cockroach and mouse droppings, but found nothing useful.

Elsie had one page of the court reporter's notes in her files. The note was brief, but no one was able to read the outdated shorthand. I called the court reporter's son in Phoenix, who was also a court reporter, but he could not transcribe his father's notes. He suggested I try a place in England where a few reporters still used the old style of shorthand. As luck would have it, an old friend of mine was in London at the time. He found a reporter who could transcribe the notes. The transcription was not a vital piece of evidence, but it was a start, and a forewarning that no information on this case would come to me easily.

Documents that should have been in the files were missing, and others took ages to find. I searched files in the Graham County Courthouse and the old Armory, but, because I had no legal training, I sometimes failed to recognize the significance of a document when I found it. On several occasions, I threw up my hands and quit. More than once, I put the work aside for at least a year, but something always drew me back. I haunted the University of Arizona law library. With the help of a patient librarian and some friendly attorneys, I worked my way through the legal maze. As the years passed, my office filled with bits and pieces of the puzzle. I began to feel there was something deliberate about all the missing papers. One problem lodged at the back of my mind and nagged me continuously. I knew that Deputy Marshal Frank Haynes had perjured himself at the inquest, and I could prove it, but I didn't know why he had lied. I told myself it wasn't important, probably just a misunderstanding. No matter how many excuses occurred to me, the problem wouldn't go away. Then one morning I woke at 4:30 AM and sat straight up in bed. That "insignificant" little lie, which had tortured me for years, was the key to everything.

Suddenly, all the pieces fell neatly into place. The picture was clear as mountain dew, and this book tells the story as I now see it.

*Barbara Brooks Wolfe*

# 1

LATE IN THE NIGHT OF FEBRUARY 9, 1918, a fierce wind whistled down the icy slopes of the Galiuro Mountains in southeastern Arizona. Four mounted men, shivering in the bone-numbing cold, rode single file along a narrow trail. All that night they endured a flesh-slashing snow that obscured their vision. Twice they lost their way. Then, just before dawn, they crossed over a high ridge. Deep in the canyon below lay the Power cabin and gold mine.

The men rode silently down a steep, switchback trail behind the house. On the slope near them, two belled horses marked with the Power brand roamed free. Halfway down the canyon, the men dismounted. One man, studying the layout of the cabin below, stepped forward. "Mart, you and Kane guard the south corners. Me and Frank will cover the north end."

A second man nodded. "That'll do it. If they try to escape through the doors or windows, we'll catch 'em in a crossfire."

The sun rose in the Galiuros at 7:07 AM on February tenth. Heavy snow clouds blocked the early morning rays. Dull grey light washed over the peaks, but deep in the canyons darkness lingered.

Inside the cabin, Jeff Power rolled out of bed in his long johns. Shivering in the cold, he tossed a log on the still-smoldering ashes in the fireplace.

Tom Power, Jeff's son, opened one eye. He watched his older brother, John, already up and dressed, light a kerosene lamp on the table. It was Sunday, John's turn to cook. Tom tugged on the covers and closed his eyes again. Across the room, Sisson, the hired hand, slept undisturbed.

Outside, a dog growled deep in her throat. Two horses thundered by the cabin, neck bells clanging.

"Mountain lion," Jeff grumbled. He grabbed a rifle and opened the canvas door.

"Throw up your hands! Throw up your hands!" a nervous voice shouted. A shot rang out.

Jeff spun around in the doorway and fell on his back in the yard.

John rushed to the door and into a hail of bullets. He ducked back and snatched a gun. With the back of his hand, he wiped blood from his eyes and opened the door a crack. He fired first to his right, and then to his left.

Tom leaped from his bed. A ball crashed through the east window and passed so close he felt the heat on his neck. The south window shattered. Glass splinters pierced his left eye. Blinded by pain, he shoved his gun through the broken window and fired. He saw the faint shadow of a man drop to the ground.

Farther up the canyon, Ed Knothe, a miner wakened by the gunfire, sat straight up in bed. His heart pounded hard against his ribs. A second burst echoed through the canyon. The shots came fast—rifles, pistols, all at once. A look of terror distorted his features. America was at war with his homeland. He'd seen hatred in the eyes of strangers when they heard his German accent. For that reason, Ed avoided conversation in general and the subject of politics in particular, but he was always afraid.

The shooting stopped. The eerie silence was more chilling than the explosion of gunpowder. The sound had come from below. Knothe tried to convince himself it was only the Murdocks, out target shooting.

At the Murdock camp, four men gathered around an open fire: Jay Murdock; his father, John; young Henry Allen; and a cowboy from the Mule Shoe Ranch. As they filled their tin cups with strong, hot coffee, they anxiously debated the wisdom of investigating the source of the gunfire.

Ed Knothe came running down the hill shouting, "Vas you target shooting?"

Jay shook his head. "The shots came from the Power place."

Inside the Power cabin, Tom held his breath and waited for the next volley. A deadly silence enveloped him like a shroud. He glanced at his brother. John sat crumpled against the doorjamb, his face a mask of blood.

Tom gasped at the sight. "You hurt bad?" he asked.

"They got a piece of my nose, and my left eye's full of wood splinters."

"They got my left eye, too. Glass. Hurts like hell," Tom said.

Sisson poked his head out from under the bed, a confused, incoherent, frightened man.

"Come on out," Tom said. "I think we got 'em all."

They heard Jeff struggling for breath in the yard. Tom opened the door a crack, waited, then pushed it aside. John followed him into the yard. Kneeling in the snow beside their father, they tried to stop the blood that gushed from the hole in his chest. Jeff was going in and out of consciousness. They moved him inside by the fireplace, but he wouldn't stay there. Out of his mind with pain, Jeff pushed his sons aside, stumbled out the door, and passed out in the yard. They made a pallet for him under the trees. When he woke, he staggered around the yard until he passed out again. Over and over, he repeated the cycle until he was finally too weak to move.

John and Tom, in agony from their eye wounds, searched the yard for their attackers. They found two bodies at the south corner of the cabin. Tom knew one was the man he had shot at through the south window. He recognized his old nemesis, Kane Wootan. The second body neither he nor John could identify. At the north corner, another corpse stared up at them through one unseeing eye. Tom shuddered and looked away, but something drew his gaze back to the dead man. He glanced at his brother. Tom knew John had also identified the third man as Graham County Sheriff Frank McBride. What they didn't know was that a fourth man had also been there that day.

Tom, John, and Sisson decided they had few choices. They could stay and face a lynch mob, ride out and surrender to the army on the border, or cross into Mexico. The fugitives headed south.

# 2

THE SHOOTOUT AT THE POWER CABIN was destined to haunt the State for generations to come. More deadly than Wyatt Earp's gunfight at the OK Corral fifty years earlier, this battle sparked a bitter controversy that would not die. Born from a lethal mixture of pride, prejudice, religion, and politics, it was a long time in the making.

The stage was set in the 1870s, following the Civil War. At that time, Graham County covered more than six thousand square miles of mineral-rich mountains, fertile valleys, and desert. The Gila River divided the county roughly in half, flowing east to west with water that was the lifeblood of southern Arizona. The eastern portion that lay south of the Gila was purchased from Mexico during Lincoln's administration. Two Apache chiefs, Cochise and Geronimo, had called this land home, and white settlers were few.

When members of the Church of Jesus Christ of Latter-day Saints were driven from their homes in the midwest, many settled along the river in the Gila Valley. Most were farmers. They established a political power base in Graham County as protection against persecution.

Outside the valley, in the desert and mountain regions, cattle grazed. Most of the ranchers were former Texans. Rugged and independent, they lived in an uneasy truce with the pious and clannish Mormons.

When the Power family arrived in Graham County in 1909, Arizona had not yet achieved statehood. The Indian Wars were over, and the most dangerous outlaws had been hanged, shot, or driven out. Law and order prevailed, more or less, but men still carried rifles in their saddle boot and six-guns on their hip. Roads were little more than wagon ruts and cow trails, but opportunities abounded. The three C's—copper, cattle, and crops—formed the basis for economic development, but the scarcity of water, oppressive heat, and poisonous reptiles discouraged settlers.

Thomas Jefferson Power, Texas born and bred, was not a man easily discouraged. He was in his early forties when he arrived in the Arizona Territory. His broad-brimmed hat hid a receding hairline. A man of average height and weight, his ordinary appearance gave no clue to the extraordinary determination, ambition, and optimism that drove him. If hard work alone could guarantee financial success, Jeff would have arrived with a million dollars in his tattered pockets. But if ever a man was born under a malevolent star, it was Jeff Power. He had little or no formal education. Twice in New Mexico he'd established a modest empire in land and cattle, and twice he was ruined by flood or drought.

He was only twenty-eight when he was widowed. His wife, Mattie Morgan, had gone with his mother to take a housewarming gift to a neighbor. While they were there, the roof on the new home collapsed and buried both women under a pile of debris. Jeff dug through the rubble with his bare hands. His mother survived, but Mattie, struck by the ridgepole, was dead. Jeff, left to support four small children and his widowed mother, traveled far and wide to find work. In Doubtful Canyon, he built a house for his family. When the work was finished, their home promptly burned to the ground.

Despite Jeff's long list of misfortunes, he remained the eternal optimist. After every tragedy, he picked himself up, put on his sweat-stained hat, and went to work again. Everyone in the Power family had that same mindset. They all worked like mules and were twice as stubborn.

Eventually, Jeff followed his eldest son, Charlie, into Aravaipa Canyon, a remote, rainbow-colored gorge between the Santa Teresa and Galiuro Mountains on the western edge of Graham County. Sandstone cliffs, carved by eons of wind and rain, towered above a wide, crystal stream. In the summer, a gentle breeze drifted through the sycamore trees, and June bugs buzzed through the willows on iridescent wings.

In the small village of Klondyke, an adobe building served as post office, supply depot, dance hall, and meeting place for the few ranchers and miners that inhabited the area.

Charlie Power was drawn to the canyon by the proximity of his mother's brothers, the Morgan boys, who owned ranches in the area. He found the peace and beauty of the canyon almost irresistible, but he knew winter storms would flood the gorge with raging water. Charlie wisely chose to settle on higher ground in the Galiuros at Rattlesnake Springs. Some say the serpent's name was a portent of evil to come, but in the wild, primitive beauty of the place, the idea seemed ludicrous. The fragrance of pine and juniper spiced the air, and high in the trees colorful birds trilled a primeval tune.

When the Power family moved in with Charlie, the whole area pulsated to the beat of pounding hammers. The men built more rooms on the cabin. They worked from dawn until dark.

At night, on a spit over the campfire, Granny cooked dinner in a big, cast-iron pot. Below the spit, a Dutch oven nestled in the coals. Soon, the tantalizing aroma of boiled beans, salt pork, and cornbread drew the men to the fire.

Jeff's youngest child and only girl, Ola Mae, fifteen years old, helped Granny Power shovel out a large area for a vegetable garden. Ola had inherited Granny's green thumb, and their garden grew so lush that folks called their home "Power's Garden Place."

Shared hardships and tragedies had molded the Power family into a tightly-knit unit. Together, the single-minded family seemed able to accomplish almost anything. Jeff worked tirelessly and expected everyone to do the same.

Charlie, the eldest son and Jeff's right-hand man, had great physical strength. The square line of his jaw testified to his proud and stubborn nature. It was Charlie who looked after the family during Jeff's long absences in New Mexico. He was only two years older than John, but both Tom and John looked up to Charlie and didn't mind taking orders from him.

John was the best marksman in the family, and he had a way with animals. Quiet, shy, and sentimental, John considered his animals part of his human family. His horse followed him around the yard like a dog.

Tom, a handsome, outgoing young man with a wry sense of humor and a gift of gab, was a jack-of-all-trades. He took on any job assigned to him and did it well.

Ola, fifteen years old, was everybody's favorite. Dark hair parted down the middle and rolled up neatly on both sides framed her sweet, round, slightly impish face. Tall for her age, her slender body moved gracefully. Her brothers teased her, of course, even roped and tied her on occasion, but she could hold her own with them, and no one else dared bother her when they were around. On the family's yearly shopping trip to the valley, Tom and John often gave up something they wanted so Ola could have some girlie thing that caught her fancy. The sight of their sister, sweet and pretty, clean and neat, made the boys swell with pride. They loved to spoil her.

Granny Power raised Jeff's children with love and understanding, but she tolerated no nonsense from anybody. Even Jeff walked softly around his mother. He remembered a day in Texas when she drove his father from a saloon with a horse whip. He watched his father scramble up a tree to escape her wrath, but his mother chopped it down with an axe. This incident left a lasting impression on her son.

In the Aravaipa community, cattlemen greeted newcomers like the Powers with mixed emotions. On one hand, they were grateful for an additional source of labor at roundup time. On the other hand, new folks brought more cattle, and that meant less grazing space in the forest reserve for herds already there. Rustling had become a way of life in the unfenced reserve. All up and down the Aravaipa, neighbors stole from neighbors. Even so, ranchers seemed to prefer the devil they knew to a newcomer.

It wasn't long before the Powers clashed with the Wootans. Each family accused the other of rustling. The Powers were outnumbered. "Shake a tree in the Aravaipa, and a Wootan falls out," the natives would say. The Powers had moved so often, they were accustomed to the problems that beset outsiders. Tom, John, and Charlie learned about safety in numbers the hard way. The three stuck together. They didn't go looking for trouble, but they didn't run from it, either.

Charlie soon made it clear to everybody that he wouldn't tolerate a bully, and he wouldn't back down from a fight. At roundup time, Charlie, all muscle and sinew, had no trouble finding a job. An old-timer who remembered him said, "Ol' Charlie was still in his teens when the family first come here. Some of the cowhands decided to have a little fun at the kid's expense. One ol' boy said he'd separated Charlie's horse from the herd on account of the animal havin' distemper. Charlie knew the horse wasn't sick, and he took it as an insult. He told the man to put the horse back with the herd. When the cowboy refused, Charlie drew his gun. The startled man threw up his hands and said, 'You wouldn't shoot an unarmed man, would you?' Charlie said, 'Nope,' and tossed him a gun. Well, the old hand made his peace with the kid, but after that nobody messed with Charlie."

Not everyone in the canyon was inhospitable to the new arrivals. The Morgan brothers welcomed their Power relatives, of course. The Claridges and the Habys did what they could to help them, and these were valuable friends to have. They were well liked and respected by others in the area. Gregory Haby and his wife, Anna, were born in Texas of German parents. They spoke German in their home, and their English was heavily accented. Gregory, a short, round bull of a man with a luxuriant handlebar mustache, also spoke Spanish, which made him popular with the earliest settlers in the canyon, like the Salazars.

The Wootans and Powers continued to argue over grazing rights and missing cattle. Who stole what from whom is still a mystery. Ray Claridge, who knew both families, said, "All I know is the Powers always ended up with fewer steers than they started with."

The friction between the Powers and Wootans was common knowledge. If some suspected the Powers of rustling, others believed the Wootans hoped to drive the Powers out and claim their grazing rights.

As time passed, Charlie tired of the constant problems. He sold his ranch to his family and returned to New Mexico. Some say Jeff and Charlie had an argument and came to blows. The Powers denied it. Tom said his brother simply had a belly full of the hard life they led. In any case, Charlie's departure could not have come at a worse time for Jeff. A man who owned a gold mine in the Galiuros had died in a gunfight over a lady's reputation. The mine had reportedly produced $65,000 in gold during the previous year. Jeff, the incurable optimist, believed his wildest dreams were about to come true, but he needed all the manpower he could muster.

It did seem as if Lady Luck had actually smiled on Jeff Power at last. He sold his cattle and bought the mine from the dead man's estate. He even found a man to replace Charlie: Tom Sisson, a veteran of the Indian Wars on parole from State Prison. Willing to work for food, shelter, and promises, he was the answer to Jeff's prayers. The two men, who were about the same age, became friends.

Although a jury had convicted Sisson of horse theft, Jeff never believed he was guilty of the crime. Sisson had not tried to hide the animal or change the brand, and he served only thirteen months before he was paroled for good behavior. Sisson, a former army scout, claimed the horse belonged to his partner, who was killed in a brawl with a Mexican. After his partner's death, Sisson, who could neither read nor write, made no attempt to find the bill of sale. During his trial, he discovered his partner had bought the horse from an Apache without the knowledge of the Indian Agent. Since the horse belonged to the tribe, the sale was not legal. This story, if true, seemed to indicate Sisson's karma was as bad as Jeff's, and Lady Luck didn't smile on their affiliation; she snickered.

Soon after Sisson joined the family, the Powers moved to temporary quarters in Klondyke. Jeff had big plans for the mine. He needed a boiler, and he wanted a stamp mill to replace the old arastra.[1] Although men had mined in the Galiuros for years, there were no roads to transport machinery and ore. Jeff Power set out to remedy that situation. Starting at Habys' ranch, Jeff, his two sons, and Sisson planned to clear a path twenty-five miles long over rugged mountain terrain. It was a monumental job most folks considered impossible, but the road was actually progressing when tragedy struck again.

One day, Ola and Granny packed a picnic lunch, borrowed a horse, hitched it to a buggy and rode out along the new road to visit the men. When they returned to their home in Klondyke, the horse balked at the unfamiliar gate. Prodded by Ola, the frightened animal reared, then raced around the yard, dragging the buggy over rocks and stumps until it overturned. Ola's scrapes and bruises were painful, but Granny's bones were old and brittle. She died of her injuries. Her death left a gaping hole in everyone's heart, but no one felt the loss more than Ola. Granny was the only mother she had ever known and her only female companion.

Without her grandmother, Ola grew desperately lonely. Sitting in Granny's old rocker with Tomba the cat in her lap, she recalled the old lady's wit and wisdom. Memories of small things, like Granny's passion for dried prunes, brought tears to her eyes. Inevitably, Ola's thoughts turned to her grandmother's faith in the afterlife. Oh, how she'd scoffed at Jeff's doubts.

---

[1] A saucer-shaped, concrete dish for grinding ore. A sweep attached to the grindstone is usually powered by a burro.

Ola could still hear her say, "Jeff says he don't believe in God, but he prays loud and long when he's sick." Remembering, she smiled a little.

They buried Granny high on a mesa in the Klondyke cemetery. Beside her grave, a determined cactus spread deep roots in the rocky soil. Thorns as sharp as Granny's tongue lay hidden behind shimmering red blossoms. The lusty cactus seemed a fitting tribute to Martha Jane Power.

After the funeral, Ola lived alone for the first time in her young life. Jeff moved the family into an abandoned mining camp on the mountain. The men gave a two-room log cabin to Ola, and they slept outside in the ruins of an old saloon.

Jeff, Sisson, and the boys worked long hours on the road. All day, Ola cooked, cleaned, and rubbed her fingers raw on a washboard. Lonely and grieving for her grandmother, she

worked without complaint. Ola loved her family and knew their future depended on the success of the mine.

The men did what they could to make her life a little easier. Every morning, Jeff woke early to cook breakfast. "Let Baby sleep a little longer," he whispered to the boys. At night, after dinner, the men brought out a worn deck of cards and challenged Ola to a game of Pitch.

Sometimes, John, the quiet brother, took down Ola's fiddle and begged her to play for them. Music, Ola decided, brought out the best in John. One stroke of the bow, and his face lit up the cabin. No matter how tired they were, the men stamped their boots and clapped their hands to the music. Tom, the show-off, even did back flips off a bench.

When the dishes were done and the last man gone, Ola fell into bed exhausted. A pine-scented breeze drifted through the window, and the moon bathed the room in golden light. Tomba, purring softly, snuggled against her breast. Whatever dreams Ola dreamed, whatever hopes filled her heart, they were not to be. Fate had her own plans for Ola Mae Power.

# 3

In the Galiuros, day after day, month after month, the Powers followed the same routine. They rose early and worked until dark on the road, oblivious to the world beyond them. The family had little time for politics and couldn't know the election of 1916 was about to change their lives forever.

In 1912 Arizona became the last of the contiguous states admitted to the Union. By 1916 the state eagerly anticipated the second general election in her brief history. Anxious to overcome her "wild west" image and attract new settlers to the state, Arizona's candidates left their six-guns at home and shot from the lip.

George W. P. Hunt, a Democrat, was the incumbent candidate for Governor. A balding walrus of a man, he ran against Tom Campbell, a tall, good-looking young Republican. The Secretary of State declared Campbell the winner, but Hunt refused to vacate his office in the Capitol Building. In January 1917, Campbell was formally inaugurated, and Hunt took the oath of office in a ceremony he arranged for himself. Both men delivered a State of the State address to the legislature, and both later signed checks for government expenses. When the State Treasurer, a Democrat, refused to honor Campbell's signature, the frustrated Republican appealed to the State Supreme Court. The court tried to oust Hunt, but he filed a writ of mandamus and demanded a recount. For eleven months, the two warring governors ruled Arizona. Eventually, a judge invalidated the ballots from one district, whereby Campbell lost 116 votes, and Hunt won the election by thirty-nine. Hunt's pet project also won. As President of the Anti-Capital Punishment Society of America, Hunt was a strong advocate of penal reform. In 1912 Arizona voters had refused to abolish what Hunt called "legalized murder," but in 1916 capital punishment was abolished in Arizona by a slim margin of 152 votes.[2]

In Graham County, E. L. Spriggs ran for County Attorney against W. R. Chambers, a well-educated, ambitious lawyer who hailed from Danville, Illinois. Like most voters in Graham County, Spriggs was both a Democrat and a Mormon. Chambers was neither, but he took a strong stand in favor of prohibition, while Spriggs made a living defending bootleggers. Chambers won the election in a landslide.

The race for Sheriff was a contest between Democrat Frank McBride and Republican Brig Stewart. McBride, a former Mormon missionary, was in his early forties and the father of seven children. The wiry, square-jawed man had worked as a farmer, cow-puncher, carpenter, and deputy sheriff. McBride, who was well connected through his church, supported strict

[2] Hunt's belated victory in 1916 left a bad taste in the mouths of his critics, but Hunt went on to serve seven intermittent and controversial terms as Governor. Democrats, proud of his popularity, like to cite the number of men who voted for Hunt after he was dead. Republicans said that was fair, considering the number of dead men who voted for Hunt while he was alive.

prohibition laws and was touted as a clean-living, impartial candidate.

Tom Power supported McBride's candidacy, but neither he nor John voted for him—or for anyone else. Kane Wootan, the registrar in Klondyke, barred them from the polls on grounds of illiteracy. Admittedly, the Power brothers' education had been limited and sporadic, but that was not so unusual in that place and time, and both Tom and John could read and write. The Powers would not soon forget the humiliation they suffered at Wootan's hands.

In November, McBride won the election. He chose a fellow churchman, Martin Kempton, as Undersheriff. Kane Wootan became a deputy under McBride. True to his campaign promise, McBride declared war on bootleggers, a cause enthusiastically supported by his teetotaling Mormon constituency.

According to a popular story, smugglers hauling liquor from New Mexico traveled the road that ran by Undersheriff Kempton's farm. A car with a heavy load made a distinctive sound bouncing over the ruts. Mrs. Kempton, a light sleeper with an educated ear, roused her husband at the first telltale sound. The Undersheriff arrested the culprits and confiscated the forbidden elixir. When the courthouse basement bulged with contraband, the water wagon sprinkled Safford's dusty streets with bathtub gin—or so the story goes.

While the Sheriff battled bootleggers in Graham County, most of Europe was embroiled in a full-fledged war with Germany. Mexico, on Arizona's southern border, was in the throes of a revolution. For nearly a year, the State militia had guarded Arizona's border against Pancho Villa's raids. In 1916 General "Black Jack" Pershing marched his army into Mexico, uninvited, in a futile attempt to capture Villa. This violation of Mexican sovereignty created a breach between the United States and Mexico.

An official letter, sent secretly by Germany's Secretary of Foreign Affairs to Germany's Minister in Mexico, directed him to attempt to unite Mexico and Japan with Germany in a war against the United States, if America should declare war on Germany. In exchange, Germany would offer to return to Mexican rule much of the American southwest. Mass paranoia gripped Arizona.

President Wilson asked Congress to pass a draft law, but conscription was never very popular with voters. The politicos argued that enough men would volunteer in the state militias, which could then be federalized as needed. Governor Campbell called for volunteers in Arizona's "Fighting First Infantry." In Company B, commanded by Graham County officers, fifteen men refused to take the federal oath, and so it went throughout the nation. Congress, finally recognizing the need for a draft, passed a conscription law in May 1917.

In Safford, the Graham County seat, the topic of conversation shifted from bootleggers to war. Men exchanged rumors in front of the bank, at the feed store, and in the lumberyard. Woman grouped together at Relief Society, quilting bees, and afternoon socials. Rumors spread through the town like a brush fire after a five-year drought.

Back in the Galiuros, oblivious to the mounting storm, the Powers moved into an abandoned two-room frame house near the mine in Kielberg Canyon. Two years of backbreaking labor had very nearly accomplished the impossible. Only a few miles remained to complete the road. Jeff Power, now in his early fifties, had achieved what no other man had dared attempt. The road was narrow and too steep for an automobile—a forty percent grade in some places—but a good team of horses could haul materials to and from the mine. With the end in sight, the men worked with renewed energy.

Around the end of May, Ola complained of a sore throat and fever. She and Jeff rode twenty-five miles down the mountain to the Haby Ranch where Jeff stored his automobile. They drove another fifty miles to Safford to see Dr. Warner. The doctor diagnosed Ola's illness as tonsillitis and prescribed medication. On the return trip Jeff picked up a copy of *The Graham County Guardian*, Safford's weekly newspaper. Ten miles west of Safford the car broke down. Forced to wait several days for repairs, Jeff and Ola stayed with the Johnson family in Pima. Years later, Mrs. Johnson described their visit.

"My husband had stayed with the Powers a time or two up in the Galiuros. When Jeff's car broke down in Pima, we were happy to return the favor. Ola didn't feel good, but she wasn't any trouble. Neat as a pin, that's the kind of girl she was. Just took her medicine and hung around with my little sister."

On June fourth, Ola and Jeff returned home with a week-old copy of *The Graham County Guardian*. According to Tom Power, it was from this paper they first learned of the war. They also learned that men between the ages of twenty-one and thirty were required to register for the draft on the following day. Tom and John, both in their mid-twenties, failed to register.

Both Power boys later claimed they tried to sign up at the Post Office in Redington, an isolated village on the opposite side of the mountain from Klondyke. Redington, closer to the mine than Klondyke, was the Powers' mailing address. According to Tom and John, the Postmaster told them the army wanted only volunteers. He allegedly advised them the Forest Ranger would notify them if and when they were needed.

Others claimed Jeff considered the conflict "Europe's war" and felt that no American was obligated to fight on foreign soil. Some say Jeff, still fuming over Wootan's barring his sons from the polls, declared, "If my boys ain't good enough to vote, they ain't good enough to fight." The Power boys denied these stories, but there is only one certainty: John and Tom Power did not register for the draft, a misdemeanor under the conscription law of 1917.

That summer the sons of Graham County marched down Safford's Main Street under waving flags. Cheering crowds lined the road, and the band played "Over There" all the way to the train depot. Proud fathers with chins trembling shook hands with departing sons. Teary-eyed mothers and sweethearts kissed the boys goodbye. Aboard the train, young recruits rushed to the windows for a last glimpse of home, but Tom and John Power were not among them.

# 4

IN THE SUMMER OF 1917, ARIZONA was in crisis. The tensions of war, the oppressive heat, and the confusion caused by two ruling governors pushed tempers to the boiling point.

The Industrial Workers of the World, a socialist labor union opposed to the war, was a menace that loomed large in southern Arizona mines. Union members, known as "Wobblies," threatened to halt production of copper essential to the war effort.

The hysterical editor of *The Graham County Guardian* warned readers that Wobblies would be tarred, feathered, and driven out of the county. "Every Arizonan," he wrote, "should constitute himself a committee of one to deal with sedition and utter lack of patriotism. He or she should report every disloyal utterance and such charges should be supported by affidavits and submitted to the Council on Defense." Soon the Graham County Courthouse files bulged with affidavits from neighbors reporting neighbors.

Bisbee, a mining town in Cochise County, exploded in violence. An angry mob herded 1,200 alleged Wobblies into stifling boxcars and shipped them into the desert without food or water.

Men not in uniform needed a good excuse. Some who worked on farms and ranches were eligible for exemption from military service. A few abused the privilege, but even those with legitimate reasons drew harsh criticism. News that the Power boys had failed to register spread through the Gila Valley like a deadly plague. Every day the fever grew.

In Klondyke, most believed the Powers did not understand the seriousness of their offense. In Safford, folks weren't as charitable. Few in town knew the Power family, but their imagination worked overtime. According to rumor, the Power boys were not just draft dodgers or "slackers," they were German sympathizers, traitors, hiding in the mountains to avoid the law. Other German sympathizers supposedly supplied them with food and other necessities. Some even claimed the Power name was German.[3]

Summer faded into fall. In the Galiuros, warm days turned crisp around the edges. Jeff replaced the antiquated arastra with a stamp mill. They needed only a boiler to reap their mine's golden harvest. The men were filled with a glorious confidence so palpable it reached out and grabbed Ola Mae. On November twenty-ninth, a jubilant Ola greeted her twenty-third birthday with mounting optimism. Not a single cloud appeared on the horizon, and her whole world took on a new luster.

A week later, while Jeff and Sisson were at work in the blacksmith shed, Ola made several trips to the shed for canned goods. After she called the men to supper, Jeff found her writhing in convulsions on the bed. He clung to her twisting body and shouted at Sisson to ride for help. His calloused hand stroked Ola's brow. Again and again, Jeff called her name. Ola's eyelids fluttered. She whispered, "Poison," and died in his arms.

[3] The first record of the Power name was found in Ireland.

Sisson raced some fourteen miles to the nearest ranch. Hours later, he returned to the cabin with Joe Boscoe and his wife, Mel. The Power men, out of their minds with grief, were inconsolable. Nothing anyone said or did penetrated their misery. Unable to comfort his friends, Sisson left the Boscoes at the cabin and rode down the mountain to buy a coffin.

Mel and Joe Boscoe had only recently moved to the area. They didn't know the Powers, but they'd heard disturbing rumors. Mel longed to flee, but the agony etched on Jeff's ashen face tugged at her heart. When he asked her to prepare Ola for burial, Mel's whole body recoiled. The stench of death sickened her. Never before had she touched a dead body, yet she couldn't refuse the suffering man. Shuddering, she rolled up her sleeves and filled a basin with water. Joe Boscoe, seeing the dread in her eyes, stood beside her. Together they bathed the body.

The feel of cold flesh ran through Mel's fingers like an electric shock. Working with desperate speed, she tried not to look at Ola's face. The girl was so young and pretty.

Fumbling through Ola's wardrobe, Jeff chose a blue silk dress and laid it on the bed.

Mel's voice quivered. "Shall I cut it up the back?"

Too dazed to think clearly, Joe shook his head. He slipped his arm under the corpse and lifted Ola's shoulders. Suddenly, her head snapped back. Mel, barely able to stifle a scream, gasped. Her body trembled and sweat dripped from her forehead. She tugged at the dress. The blue silk slid over Ola's dark hair. Quickly gathering a sleeve in her hand, Mel reached for Ola's arm. It was rigid, immovable, locked against her chest. Frantic, Mel glanced at her husband. Her eyes brimmed with tears.

Joe tugged at the arm with all his might, but rigor mortis had set in. His struggle with the arm seemed endless, but eventually both sleeves were in place, and he collapsed in a chair, exhausted.

All night, a terrified Mel Boscoe sat with the body and dabbed at a dark liquid that oozed from Ola's mouth. Outside, under a cold witch's moon, Tomba scratched at the roof and howled for his mistress. The mournful cry tore at Mel's ragged nerves.

Around midnight, Sisson arrived at the Haby Ranch in Klondyke. The clatter of hooves in the yard woke Anna Haby. She ran to the window and called to him. Sisson shouted something about a coffin, leaped into Jeff's car, and disappeared in a cloud of dust.

Early the next morning, Sisson arrived at the funeral parlor in Safford. When the news reached the courthouse, Sheriff McBride misunderstood it was Jeff Power who had died. McBride knew the Power boys had not registered for the draft, and he'd heard rumors about their alleged German sympathies. This was his opportunity to investigate the family, and he jumped at the chance. Accompanied by Dr. W. E. Platt, the Sheriff sped down the valley, over the mesa, and through the pass to Klondyke.

Meanwhile, the Powers had moved Ola's body to Habys' Ranch, where they waited for Sisson to arrive with the coffin. The Habys, stunned by Ola's sudden death, shared the family's anguish.

When Sheriff McBride and Dr. Platt arrived at Klondyke, they learned the body was at Habys'. The Powers' friendship with the German-speaking Habys may have fueled the Sheriff's suspicion, but if he doubted the Habys' loyalty to America, he didn't dare say so in Klondyke. The Habys' son, Romeo, was an officer in the U.S. Army.

Expecting to find Jeff Power's corpse at the ranch, McBride and Platt were shocked to discover the body of a young woman.

The corpse was black from the breast up. Dr. Platt, noting the slight tilt of Ola's head, announced her neck was broken. Then he spotted her distended abdomen and leaped to the conclusion she'd died of a botched abortion.

Outraged, Anna Haby snapped, "You didn't know her." When Joe Bleak, ex officio coroner, called for an autopsy, Anna refused to leave Dr. Platt alone with the body. She followed him into the bedroom.

Dr. Platt removed the upper part of Ola's spine. He found one dislocated vertebra but no fracture. Further examination proved she had not been pregnant. Her hymen was intact. Ola's organs appeared normal except for a few tiny white spots on her stomach tissue. Dr. Platt was mystified.

At the inquest, Joe Boscoe testified he saw no bruises or other marks of violence on Ola's body.

Dr. Platt stated he did not know the cause of Ola's death. The jury's verdict, "Death from an unknown cause," satisfied no one. Dr. Platt sent Ola's organs to Tucson for analysis. Laboratory tests revealed no trace of poison.

The mystery surrounding Ola's tragic death invited rumors and speculation. *The Graham County Guardian* reported her death as an apparent suicide. Joe Bleak said he'd asked Dr. Platt, "How long does a body have to lay before the skin slips like hers does?" Platt allegedly answered, "Three to five days." This indicated Ola's death occurred days before her family reported the tragedy. Other conditions, some claimed erroneously, supported this theory; the stench in the cabin, the extent of rigor mortis, and the discoloration of the body.

Ola's last word, "Poison," inspired more gossip. Strychnine was readily available in cattle country, and cyanide was used in some stages of gold processing. Both were presumably available to the Power men.

In Klondyke, speculation was more circumspect. Knowing the family's deep love for Ola, few suspected the Powers of foul play. When someone suggested Sisson as a possible suspect, one woman countered, "Tom woulda killed him 'fore he blinked."

The Power family harbored their own suspicions. They believed the Wootans had poisoned their food supply in an effort to take over their gold mine. Ola's family deeply resented the sheriff's intrusion into their grief. Platt's remarks and his obvious suspicion enraged them, but the mutilation of Ola's body was almost more than they could bear. They buried their beloved Ola next to her grandmother in the Klondyke Cemetery. She was the last of the Power women. No evidence that the Wootans poisoned Ola ever existed, but a monument to the Powers' suspicion remains. A discreetly worded inscription on Ola's headstone reads, "Poisoned by Unknown Person."

# 5

THE POWERS' MOURNING FOR OLA seemed never-ending. Platt's remarks had made a sordid scandal of her tragic death and cast suspicion on her family.

A few days after Ola's burial, the Powers' neighbor, Jay Murdock, visited the family. Jay had been away working on a fence line when Ola died. Shocked to hear of her sudden death, he went to the cabin to hear the story.

The dark room was a yawning cave of misery. John, slumped down in a corner, held Ola's violin tight against his chest. Lost in thought, he seemed not to notice Jay's presence.

Tom greeted Jay with a nod and motioned him to an empty chair. A silence of futility and gloom hung in the air. Ola's death, and the mutilation of her body, was a festering laceration on Tom's soul. He couldn't think beyond it, and he spoke bitterly of McBride.

During the conversation, Jay asked, "Suppose McBride and his posse were to walk in here?"

Tom shrugged. "Let him come. I'd just as soon die now as anytime." His eyes sparked as he added, "While they're getting me, I'll get some of them, maybe more than they'd get of us."

A few weeks later, Jay rode into Safford to have his horse's teeth filed. McBride seized the opportunity to question Jay about the Powers. Jay did not like being placed in the middle of a dispute. He thought it best not to offend either side. Jay explained to McBride that the Power boys didn't understand how failure to register was a serious crime. He also told the sheriff the Powers had other things on their minds. The grief-stricken family was obsessed with Ola's death, he said, and he repeated his conversation with Tom.

McBride wrote a letter to Jeff Power explaining the consequences of draft evasion, and he asked Jay to deliver it. Jay took the letter with him and returned to the mountain.

Jeff did not respond to McBride's letter.

The sheriff, convinced the Powers were guilty of sedition, murder, or both, hoped to take legal action against them. Two obstacles stood in his way. He had no evidence that Ola was murdered by anybody, let alone her family. He couldn't arrest the boys for draft evasion because it was a federal offense and outside his jurisdiction. Frustrated, McBride consulted with County Attorney Chambers.

After his conversation with Chambers, McBride filed a complaint against the Power boys with Justice of the Peace U. I. Paxton. Paxton issued federal warrants for John and Tom Power, charging failure to register for the draft. McBride sent the warrants to Deputy U.S. Marshal Frank Haynes in Globe, who forwarded them to U.S. Marshal J. P. Dillon in Phoenix.

Dillon, a blonde, over-worked public servant, had never heard of the Power brothers, and his budget was exhausted. The cost of sending Deputy Haynes on a long trip into a nearly inaccessible area to arrest two men for a misdemeanor sent his blood pressure soaring. The trip from Globe to Safford to the

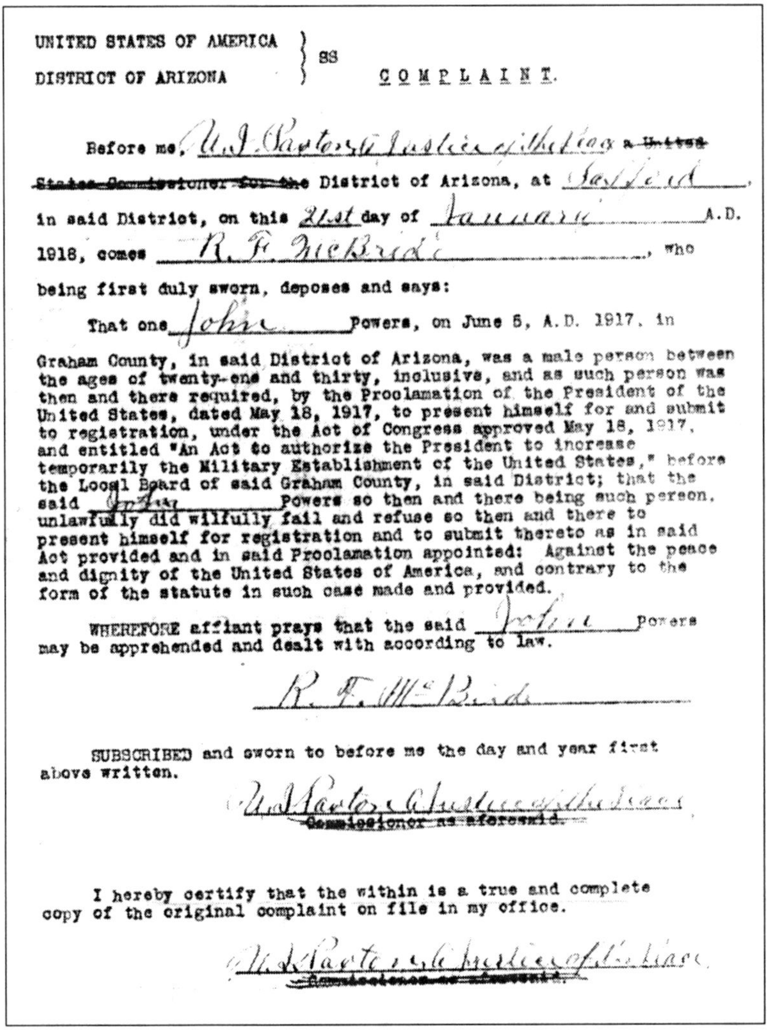

UNITED STATES OF AMERICA }
} SS
DISTRICT OF ARIZONA } COMPLAINT.

Before me, U. J. Paxton, Justice of the Peace ~~a United States Commissioner for the~~ District of Arizona, at Safford, in said District, on this 21st day of January A.D. 1918, comes R. F. McBride, who being first duly sworn, deposes and says:

That one John Powers, on June 5, A.D. 1917, in Graham County, in said District of Arizona, was a male person between the ages of twenty-one and thirty, inclusive, and as such person was then and there required, by the Proclamation of the President of the United States, dated May 18, 1917, to present himself for and submit to registration, under the Act of Congress approved May 18, 1917, and entitled "An Act to authorize the President to increase temporarily the Military Establishment of the United States," before the Local Board of said Graham County, in said District; that the said John Powers so then and there being such person, unlawfully did wilfully fail and refuse so then and there to present himself for registration and to submit thereto as in said Act provided and in said Proclamation appointed: Against the peace and dignity of the United States of America, and contrary to the form of the statute in such case made and provided.

WHEREFORE affiant prays that the said John Powers may be apprehended and dealt with according to law.

R. F. McBride

SUBSCRIBED and sworn to before me the day and year first above written.

U. J. Paxton, Justice of the Peace
~~Commissioner as aforesaid.~~

I hereby certify that the within is a true and complete copy of the original complaint on file in my office.

U. J. Paxton, Justice of the Peace
~~Commissioner as aforesaid.~~

Power mine and back again was roughly 300 miles, but cost wasn't the Marshal's only concern. In his mind, the legality of the warrants was questionable. Why hadn't McBride filed his complaint with a federal commissioner, as was customary? After all, failure to register was a federal crime, and the Justice of the Peace was not a federal officer. Did he even have the

# Warrant of Arrest

UNITED STATES OF AMERICA, } ss.
..................District of Arizona, }

**TO THE UNITED STATES MARSHAL:**

COMPLAINT upon oath having been this day made before me, a ~~United States Commissioner~~ Justice of the Peace ~~in the~~ in and for Graham County ~~District of the~~ State of Arizona, that the offense of Failure to register in violation of Sec. 5 Act of May 1917 has been committed, and accusing John Powers thereof,

YOU ARE THEREFORE commanded by the President of the United States forthwith to arrest the above-named John Powers and bring him before me forthwith at my office, in the .................. District of Arizona, or in case of my absence, or inability to act, before the nearest and most accessible Commissioner within this District.

Given under my hand this 21st day of January A. D. 1918

U. J. Paxton, a Justice of the Peace ~~U. S. Commissioner.~~

No. 362—WARRANT OF ARREST—The McNeil Company's Red-Line Blanks, Phoenix

# Warrant of Arrest

UNITED STATES OF AMERICA, } ss.
..................District of Arizona, }

**TO THE UNITED STATES MARSHAL:**

COMPLAINT upon oath having been this day made before me, a ~~United States Commissioner~~ Justice of the Peace ~~in the~~ in and for Graham County ~~District of the~~ State of Arizona, that the offense of Failure to Register in violation of Sec. 5 Act of May. 18. 1917 has been committed, and accusing Thomas Powers thereof,

YOU ARE THEREFORE commanded by the President of the United States forthwith to arrest the above-named Thomas Powers and bring him before me forthwith at my office, in the .................. District of Arizona, or in case of my absence, or inability to act, before the nearest and most accessible Commissioner within this District.

Given under my hand this 21st day of January A. D. 1918

U. J. Paxton, a Justice of the Peace ~~U. S. Commissioner.~~

No. 362—WARRANT OF ARREST—The McNeil Company's Red-Line Blanks, Phoenix authority to issue a federal warrant? Dillon took the warrants to the federal attorney for a legal opinion.

At Dillon's request, the federal attorney examined the documents. Following a time-honored bureaucratic tradition, he played it safe and simply covered his flank. The attorney demanded the elimination of all reference to federal offices on the warrants and on the complaint. He also advised Dillon to return the papers to the maker for these corrections.

On January 25, Dillon wrote to Deputy Haynes and enclosed the documents with the list of corrections demanded

AT PHOENIX—Commencing first Mondays in April and October.
AT PRESCOTT—First Mondays in March and September.

December.

DEPARTMENT OF JUSTICE.

OFFICE OF UNITED STATES MARSHAL,

DISTRICT OF ARIZONA.

PHOENIX. January 25, 1918

Frank Haynes,
Deputy U. S. Marshal,
Globe, Arizona.

Dear Sir:

Enclosed find two warrants for John and Thomas Powers. The warrants are not complete, and the United States Attorney has attached a slip of paper to each warrant showing what should be added, and requests that you have the Justice of the Peace who issued the warrants attend to it before the warrants are executed. I understand that you have been fully advised with regard to these two men. I have written the Department asking for authority to incur the necessary expense in apprehending them, and you will not make any effort towards serving these warrants until you hear further from me. If it becomes necessary for you to visit Safford before that time you might have the Justice of the Peace make the correction at that time, but it will not be necessary to make a special trip to Safford for the purpose. You can send them to him by mail and have him return them to you when they are completed.

As soon as I hear from the Department regarding my request I will advise and instruct you.

Respectfully,

J P Dillon
U. S. Marshal.

W/F.

Encls.

by the federal attorney. The Marshal explained he had asked the Justice Department for additional funds and was awaiting their reply. In his letter to Haynes he wrote, "You will not make any effort toward serving these warrants until you hear further from me."

After Dillon's request for funds was granted, the Marshal communicated with Haynes again. Dillon wrote, "…go at once

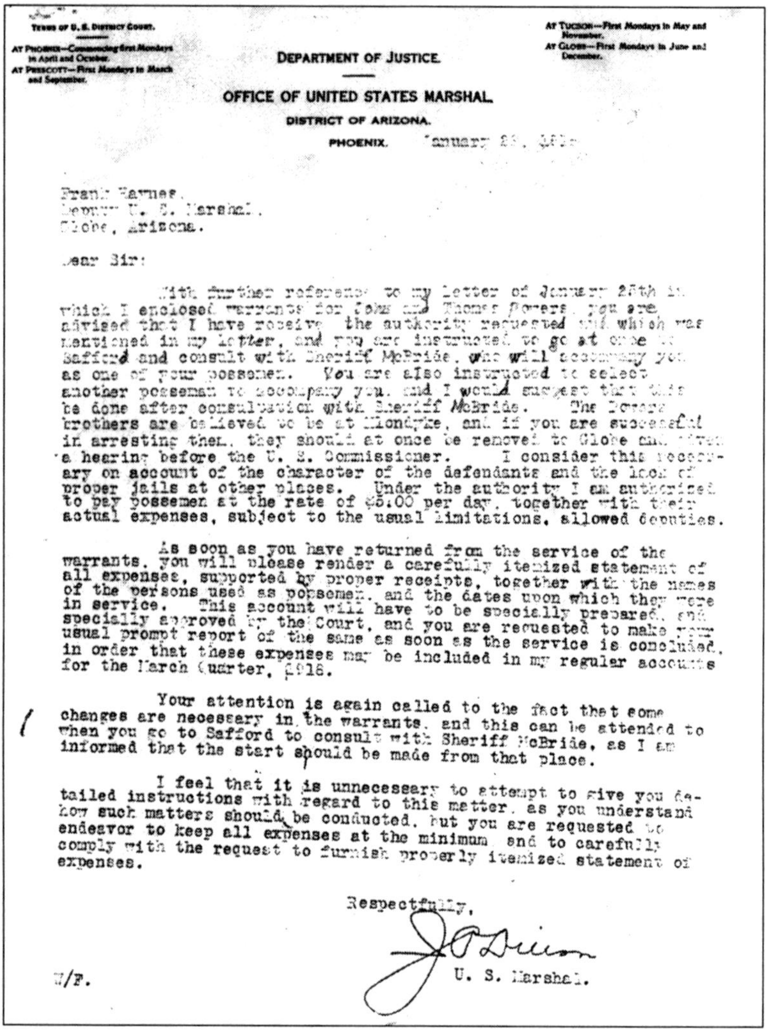

TERMS OF U. S. DISTRICT COURT.
AT PHOENIX—Commencing first Mondays in April and October.
AT PRESCOTT—First Mondays in March and September.

AT TUCSON—First Mondays in May and November.
AT GLOBE—First Mondays in June and December.

DEPARTMENT OF JUSTICE.

OFFICE OF UNITED STATES MARSHAL.

DISTRICT OF ARIZONA.

PHOENIX. January 2[illegible], 191[illegible]

Frank Haynes,
Deputy U. S. Marshal,
Globe, Arizona.

Dear Sir:

With further reference to my letter of January 25th in which I enclosed warrants for John and Thomas Powers, you are advised that I have received the authority requested and which was mentioned in my letter, and you are instructed to go at once to Safford and consult with Sheriff McBride, who will accompany you as one of your possemen. You are also instructed to select another posseman to accompany you, and I would suggest that this be done after consultation with Sheriff McBride. The Powers brothers are believed to be at Klondyke, and if you are successful in arresting them, they should at once be removed to Globe and given a hearing before the U. S. Commissioner. I consider this necessary on account of the character of the defendants and the lack of proper jails at other places. Under the authority I am authorized to pay possemen at the rate of $5.00 per day, together with their actual expenses, subject to the usual limitations, allowed deputies.

As soon as you have returned from the service of the warrants, you will please render a carefully itemized statement of all expenses, supported by proper receipts, together with the names of the persons used as possemen, and the dates upon which they were in service. This account will have to be specially prepared, and specially approved by the Court, and you are requested to make your usual prompt report of the same as soon as the service is concluded, in order that these expenses may be included in my regular accounts for the March Quarter, 1918.

Your attention is again called to the fact that some changes are necessary in the warrants, and this can be attended to when you go to Safford to consult with Sheriff McBride, as I am informed that the start should be made from that place.

I feel that it is unnecessary to attempt to give you detailed instructions with regard to this matter, as you understand how such matters should be conducted, but you are requested to endeavor to keep all expenses at the minimum and to carefully comply with the request to furnish properly itemized statement of expenses.

Respectfully,

J. P. Dillon

U. S. Marshal.

W/F.

to Safford and consult with Sheriff McBride, who will accompany you as one of your posse." He ordered Haynes to take two men and to consult with McBride on the choice of the second man. The Marshal reminded his deputy to attend to the necessary corrections on the warrants before he attempted to serve them. The rest of the letter was devoted to Dillon's monetary concerns. He expounded on the necessity for itemized statements, detailed expense reports, and proper receipts. One expense item is noteworthy apart from the sums involved. Dillon wrote, "I am authorized to pay posse men at the rate of five dollars per day together with their actual expenses, subject to the usual limitations allowed *deputies*." Dillon instructed Haynes to list the dates his posse men were *in the service of the United States government.*

On Thursday, February 7, Deputy Marshal Haynes arrived in Safford. Lean, tan, and clean-shaven, he wore high, laced boots and a wide-brimmed hat that shaded his handsome features. Ambling up the courthouse steps, he looked like Hollywood's version of a western lawman. More than one woman took notice of the Deputy Marshal when he passed.

Inside the sheriff's office, Haynes returned the warrants to McBride, along with the list of corrections demanded by the federal attorney. Haynes and the sheriff discussed the possible candidates for the second posse man. Undersheriff Martin Kempton volunteered to go, but McBride chose Kane Wootan. Despite the animosity between the Wootans and the Power family, the choice wasn't entirely unreasonable. McBride and Haynes were not familiar with the twisting network of trails in the Galiuros, and Wootan knew the territory. Unwilling to disappoint Kempton, McBride took it upon himself to take the Undersheriff along to "hold the hosses."

On Saturday, February 9, when the posse left the valley in McBride's car, Wootan, McBride, and Haynes were employees of the federal government. They were hired and would be paid by the federal government to deliver federal warrants to two men living on federal land who were charged with a federal crime. Kempton was the odd man out.

When the four men arrived in Klondyke, they drove to the Haby ranch to borrow horses for the twenty-five mile trek up the mountain to the Power cabin. The Habys were known to be hospitable and accommodating to visitors, but in this case Haby did not oblige them. The posse was heavily armed and, according to Haby, they'd been drinking. He offered to ride up the mountain himself and bring the Power boys in peacefully. The posse refused his offer, and Haby refused to lend his horses.

The posse moved on to Kane Wootan's ranch, where three horses and a mule were available. After a light meal, they collected a few extra guns and rode into a dark and frigid night. Under a black, moonless sky, four mounted men rode the glory trail to keep a date with destiny.

# 6

The next morning, following the bloody gun battle, the Power brothers and Sisson searched the yard. They saw no movement, heard no whisper of a sound. Three bodies lay in the snow. Jeff Power, mortally wounded, unconscious, and barely alive, lay bleeding beside the cabin door.

Tom, John, and Sisson struggled to grasp the horrible reality. Only a few hours earlier, their future had looked so bright. Now, here they stood dazed by the terrible suddenness of events that had changed their lives forever. Misfortune, it seemed to Tom, was a family tradition, a curse they couldn't escape.

Tom's face, cut by flying glass, looked as if he'd been mauled by a wildcat. John's features were barely visible under a mask of dried blood. Both were in excruciating pain.

They did what they could for Jeff, but they knew they couldn't stay with him. On the hill behind the cabin, they found two horses and a mule tied to trees. They never guessed there had been a fourth man there that day. Tom and John rode single file up the trail on horseback. Sisson followed them on the mule. They stopped at Murdock's camp and asked the men there to look after their father.

When they were gone, young Henry Allen, barely able to contain his excitement, stuffed his shirttail into his pants, jumped on Jay Murdock's horse and raced down to Klondyke to get help. Ignoring the trail, Henry's mount dove through the canyons at reckless speed. Before long, Henry spotted tracks in the snow. Another rider had preceded him. Like Henry, he traveled at a dangerous pace. Henry, fearing he might be mistaken for one of the fugitives, paused to tie a white handkerchief to his hat, and continued his plunge through the canyons. Late that afternoon, he arrived at a ranch in Klondyke. A well-lathered horse stood snorting and panting at the hitching post. Several men had gathered in the yard. An agitated, mud-splattered man in the middle was doing all the talking. Henry knew he'd found his mystery rider. Surprisingly, the man, Deputy U.S. Marshal Frank Haynes, wasn't aware of his posse's fate. Henry told him his men were all dead.

Around the time Henry arrived in Klondyke, Jeff Power died of his wounds. All that day, Jay Murdock had tended to Jeff and tried to make him comfortable. He sheltered him in the mine tunnel and stripped his clothes to tend his wound. When Jeff was conscious, he repeated again and again, "Why did Kane shoot me? I had my hands up beggin' for peace."

Jay, an experienced hunter, knew an entrance wound from an exit wound. He determined the bullet entered Jeff's chest and exited through his back. More importantly, Jay noticed that blood gushed from the wound when Jeff's arms were raised. The bullet had made a clear passage from front to back. When Jay lowered the man's arms, a muscle slipped over the hole and diminished the flow of blood.

As snow began to fall on the mountain again, Jeff Power died. Jay covered the bodies in the yard with blankets, rugs, and tarps he found in the cabin. All that night, he protected the bodies from predators. With a rifle at his side, he rested with open eyes. Even the sound of a falling leaf roused him. Toward morning, he dozed off and woke suddenly to find himself staring down a rifle barrel. "Don't shoot. It's me, Jay Murdock," he shouted.

"Where are the Power boys?" a gruff voice demanded.

Jay shrugged. "Headed for the border, I guess."

The vigilantes had arrived. Among them was McBride's brother, Howard, and Kempton's brother, Nate. The men, angry and bent on revenge, worried Jay. They were in no mood to be reasonable, and Jay could do nothing but watch when they descended on the cabin like locusts on a wheat stalk. They trampled evidence, looted the cabin, stole Ola's violin, and helped themselves to food, which they first tested for poison on the Powers' dogs. In the end, they tied the bodies of the posse men face down on pack horses and prepared to leave. Jay begged them to take Jeff's body, too. He even offered to furnish the horse, but one man snarled, "Hang the son-of-a-bitch from a tree and let the coyotes have him."

Late that night, the vigilantes rode into town with the bodies. Full of wild tales of the gun battle, they claimed the Powers had lain in wait for the posse in a barricaded cabin. Kane Wootan, they said, was still on his horse when a bullet struck and killed him. The most damaging story came from the vigilante who claimed McBride had been shot in the head by the murdering Power boys while he lay wounded and helpless on the ground. These inaccurate stories inflamed an already enraged public. Large groups of men rode out of town in every direction to hunt down the fugitives.

The wire service picked up the story. Reporters flocked into Safford. The tale of a shootout more deadly than Wyatt Earp's battle at the OK Corral in Tombstone was irresistible.

County Attorney W. R. Chambers, a slight, balding man in wire-rimmed glasses, watched the crowd in the street from his office window. On his desk lay a once-in-a-lifetime opportunity for a small town prosecutor, a case that could make his name a household word. Of course, there were legal obstacles to overcome, not the least of which was the fact that he did not have jurisdiction in the case.

The courthouse lights burned all night. In the basement, where the autopsies took place, Doctors Schenck, Platt, and Stratton went about their grisly job in preparation for the inquest scheduled for the following day.

Meanwhile, in Klondyke, news of the sudden violence appalled the inhabitants. The dawn attack on the Powers by a heavily armed posse had all the earmarks of an ambush. The ruthless attempt to arrest two men for a misdemeanor roused both anger and fear in the ranchers and miners of the Aravaipa. Told of Jeff Power's brutal end, one thought stayed in their minds—"There, but for the grace..."

Gregory Haby said the posse men were drunk and spoiling for a fight when they stopped at his ranch to borrow horses. Another man claimed he saw the posse at the Klondyke store, "so drunk they couldn't hit the ground with their hat," he said. Outraged relatives of McBride and Kempton vehemently denied these stories. Both families were devout Mormons, and drinking was strictly prohibited by the church.

Deputy Marshal Haynes, billed as the "only living witness to the shootout," basked in media glory. Reporters portrayed him as a dashing hero in a western saga. The publicity-seeking Haynes would become Chambers' star witness—and his worst nightmare. Surrounded by reporters thirsting for tales of derring-do, Haynes' mouth shifted into high gear while his brain remained in neutral. At first, Haynes seemed determined to avoid blame for the bungled, overblown attempt to arrest the Power boys for a misdemeanor. Interviewed by a reporter from his hometown paper, Haynes bragged that he never fired his gun during the battle. This was probably true, since evidence suggests he fled as soon as he heard the first shot explode. However, the reporters didn't know that, and the vision of the Deputy Marshal standing knee-deep in bodies with his gun holstered raised embarrassing questions. In subsequent interviews, the number of shots Haynes said he fired increased with every telling. Sadly, he had publicly complimented the first reporter on the accuracy of his story and could not later claim he'd been misquoted.

As worrisome as Haynes' loose tongue was to Chambers, it wasn't his only problem. Haynes had allowed an inexperienced sheriff to usurp authority over his posse with disastrous results, and he'd deserted his men without knowing whether they were dead or alive. Fortunately, for Chambers, the public was not yet aware of Haynes' debacle, and the county attorney had more pressing problems on his mind. The unauthorized addition of the fourth man, Kempton, added substance to the impression of an ambush, and would not sit well in some quarters. Kempton's presence had to be explained. Jeff Power's death at the hands of an old enemy, Wootan, would also have to be justified. More important to Chambers was the jurisdictional problem. McBride and Wootan were in the employ of the federal government and were killed on federal land. The federal court had jurisdiction in the case. By all accounts, Chambers was a competent lawyer. If he hoped to take jurisdiction over the case, he had to deal with that problem before the inquest on the following day.

Outside the well-lit courthouse, an angry mob crowded the street. Rumors were rampant. Men set out in every direction to search for the Power boys and Sisson. Before long, sharpshooters from all over Arizona and New Mexico had joined the fray.

# FIVE COUNTIES ON TRAIL OF MURDERERS

**Sheriff R. F. McBride, Under-Sehriff Kempton and Deputy Wootan Are Shot Down in Battle Near Aravaipa Canyon; Chief Bailey and Sheriff Miles Lead Two *Posses* from Tucson.**

The first armed resistance in Arizona to the military draft occurred Sunday morning when Sheriff R. F. McBride of Graham county, Under-Sheriff M. R. Kempton and Deputy Kane Wootan were shot and killed by three men, whom they had gone to arrest, at an abandoned mine near Klondyke, Graham county.

**Sheriffs in all the Southern counties have placed posses in the field to hunt the murderers. . L. H. Morgan of Fort Huachuca, commanding the military district, has issued orders to all army patrols to watch out for the three men.**

The killers are John Powers and Tom Powers, young men, who were wanted on a federal warrant charging them with evading the draft, and Tom Sisson, about 50 years old, and ex-soldier and reported to be an ex-convict.

**Two Posses From Tucson.**

Two posses have started from Tucson. One of them under Deputy Sherif Tom Burts has gone to Reddington, for which point the killers are believed to be heading as they went South, riding on the only trail that leads through the Galluro mountains and which debouches at that point. It is highly probable that a battle between the outlaws and the posses will occur near Reddington.

Deputy United States Marshal Haynes, stationed at Safford, went to serve the warrants, the county officers accompanying him. It is reported that the outlaw, Sisson, was badly wounded. Another roundabout report is that Haynes was wounded but he cannot be seriously injured as he has been in telephonic communica-

ported that the outlaw, Sisson, was badly wounded. Another roundabout report is that Haynes was wounded but he cannot be seriously injured as he has been in telephonic communica-tion with Don Willetts, chief deputy United States marshal, who is directing the federal forces from Tucson in the absence of United States Marshal Dillon at Phoenix.

The outlaws are heavily armed and very dangerous men, it is stated in a circular issued Monday from the United States marshal's office.

**Details Are Lacking.**

Details of the encounter are lacking in Tucson. It is supposed that the party of officers were ambushed or else that the outlaws shot without warning. The fight took place at 7:30 o'clock in the morning.

Klondyke is a small mining camp on the Northeast side of the Galluro mountains near the head of the Aravaipa valley.

Posses from five counties, including Graham, Pima, Pinal, Cochise, Santa Cruz, are scouring the roads and every resource of the state is being employed to draw the net about the desperate gang.

Sheriff Miles heads one posse from Tucson and is accompanied by Chief of Police Bailey and Cattle Inspector McKinney. Deputy Burts has in his party Deputies Archer, Cummins and Spears. They went by automobile to Reddington, which is in the San Pedro valley on the Northeast side of the Catalina mountains, and about 50 miles from Tucson. It is the headquarters of the Bayless cattle ranch.

**Brothers of "Maricopa Slim"**

John and Tom Powers are brothers said to be engaged in the cattle business.

The outlaws rode south through the mountains on a brown horse, a sorrel horse and a sorrel mule.

The first information was received in Tucson Sunday afternoon by telefrom from County Attorney W. R. Chambers of Graham. He warned that the outlaws would fight.

Burts and five men left at once. Officers were busy all night organizing the hunt by telegraph and telephone. Sheriff Wheeler of Cochise, Hall of Pinal and Earhart of Santa Cruz detailed men to the capture of the killers. Sheriff Earhart is in Tucson with two prisoners en route to the state prison and has been advised by his chief deputy that the roads are being watched.

"Rewards will be offered some time during the day, in all probability," said Deputy Marshal Willetts. "Graha mcounty probably will offer not less than $1,000, the government and the state will add to it."

**Description of Hunted.**

The descriptions of the killers is as follows:

John Powers, 23 years old, sandy complexion, smooth shaven, weight

# SHERIFF AND HIS TWO DEPUTIES MURDERED

(Continued from page one.)

about 160 pounds, 5 feet, 8 inches tall, slender.

Tom Powers, 26 years old, dark complexion, weight about 160 pounds, 5 feet 8 inches tall; slender.

Tom Sisson, about 50 years old, very dark coplexion, black mustache. Has rasping voice, about 5 feet, 11 inches tall, weighs about 180 pounds.

**Terrain of the Crime.**

The Galluros are an exceedingly rugged range running Southeast and northwest, north of and aparallel to the Catalinas. Their imposing bulk rises before the traveler as his car descends into the San Pedro valley on the way to Mammoth or Reddington.

The Aravaipa canyon is a long and narrow cut in the hills to the West of the Galluros, which reaches around back of them, opening into the Aravaipa valley on their farther side. Here are two small settlements, Aravaipa and Klondyke.

It is reported that the Powers brothers had sent word to the federal authorities at Safford that if they wanted them, they should come and get them.

**Report From Safford.**

The Citizen is in receipt of a letter from Dr. H. A. Schell, the Tucson oculist, who happened to be at

says that the Powers' brothers were located at a cabin in the mountains by the posse.

"Full particulars are not known," he continues, "but it has been learned that the father of the Powers boys was shot and possibly fatally wounded. The officers were evidently killed without warning while advancing towards the cabin. Word was brought in by Former Sheriff Haines of Gila county, now a United States deputy marshal.

"The murderers had made threats to kill any officers attempting to arrest them.

"Considerable excitement prevails at Safford.

"I heard that Sissons was out of the penitentiary on parole by Hunt, but afterwards heard that he had served his time for horse stealing and the one on paralo is a different party."

J. A. McMurren of Tucson, a relative of "Maricopa Slim" Powers, special agent of the Southern Pacific, who was killed by tramps some years ago, denied reports that the Powers brothers are related to "Maricopa Slim", although the names are the same.

The Powers brothers are described in a circular calling for their arrest, issued by United States Marshall Dillon as the brothers of "Marcopa Slim".

**Account of Shooting**

According to Hayne's account, he planned to arrest the younger Powers on the draft evasion charge the Graham county officers volunteering to accompanying him that they might investigate the death of a sister of the alleged draft evaders, who died about six weeks ago, under circumstances which Haynes said the other officers considered auspicious. Haynes said that when the officers reached the Powers cabin in the mountains, after a long horseback and automobile trip from Safford, Arizona, they found Powers, his two sons and Sisson there. Haynes said they opened fire on the officers at once, and that after many shots had been exchanged McBride, Kempton and Wootan were killed and he was wounded. He hurried back to Safford to secure aid.

While he was gone Powers' sons and Sisson took the horses and arms of the slain officers and escaped into the mountains.

The elder Powers, father of the two men alleged to be evading the draft is reported dangerously wounded in the right eye.

[Above article dated February 12, 1918 reprinted courtesy of Tucson Citizen]

[Transcript of article from Tucson Citizen, February 12, 1918.]

## POSSES FROM FIVE COUNTIES ON TRAIL OF MURDERERS

**Sheriff R. F. McBride, Under-Sehriff [sic] Kempton and Deputy Wootan Are Shot Down in Battle Near Aravaipa Canyon; Chief Bailey and Sheriff Miles Lead Two Posses from Tucson.**

The first armed resistance in Arizona to the military draft occurred Sunday morning when Sheriff R. F. McBride of Graham County, Under-Sheriff M. R. Kempton and Deputy Kane Wootan were shot and killed by three men, whom they had gone to arrest, at an abandoned mine near Klondyke, Graham county.

Sheriffs in all the southern counties have placed posses in the field to hunt the murderers. L. H. Morgan of Fort Huachuca, commanding the military district, has issued orders to all army patrols to watch out for the three men.

The killers are John Powers and Tom Powers, young men who were wanted on a federal warrant charging them with evading the draft, and Tom Sisson, about 50 years old, an ex-soldier and reported to be an ex-convict.

### Two Posses From Tucson

Two posses have started from Tucson. One of them under Deputy Sheriff Tom Burts has gone to Reddington, for which point the killers are believed to be heading as they went South, riding on the only trail that leads through the Galiuro mountains and which debouches at that point. It is highly probable that a battle between the outlaws and the posses will occur near Reddington.

Deputy United States Marshal Haynes, stationed at Safford, went to serve the warrants, the county officers accompanying him. It is reported that the outlaw, Sisson, was badly wounded.

Another roundabout report is that Haynes was wounded but he cannot be seriously injured as he has been in telephonic communication with Don Willetts, chief deputy United States marshal, who is directing the federal forces from Tucson in the absence of United States Marshal Dillon at Phoenix.

The outlaws are heavily armed and very dangerous men. It is stated in a circular issued Monday from the United States marshal's office.

### Details Are Lacking

Details of the encounter are lacking in Tucson. It is supposed that the party of officers were ambushed or else that the outlaws shot without warning. The fight took place at 7:30 o'clock in the morning.

Klondyke is a small mining camp on the Northeast side of the Galiuro mountains near the head of the Aravaipa valley.

Posses from five counties, including Graham, Pima, Pinal, Cochise, Santa Cruz, are scouring the roads and every resource of the state is being employed to draw the net about the desperate gang.

Sheriff Miles heads one posse from Tucson and is accompanied by Chief of Police Bailey and Cattle Inspector McKinney. Deputy Burts has in his party Deputies Archer, Cummins and Spears. They went by automobile to Reddington, which is in the San Pedro valley on the Northeast side of the Catalina mountains, and about 50 miles from Tucson. It is the headquarters of the Bayless cattle ranch.

### Brothers of "Maricopa Slim"

John and Tom Powers are brothers said to be engaged in the cattle business.

The outlaws rode south through the mountains on a brown horse, a sorrel horse and a sorrel mule.

The first information was received in Tucson Sunday afternoon by telegram from County Attorney W. R. Chambers of Graham. He warned that the outlaws would fight.

Burts and five men left at once. Officers were busy all night organizing the hunt by telegraph and telephone. Sheriff Wheeler of Cochise, Hall of Pinal and Earhart of Santa Cruz detailed men to the capture of the killers. Sheriff Earhart is in Tucson with two prisoners en route to the state prison and has been advised by his chief deputy that the roads are being watched.

Rewards will be offered sometime during the day. "In all probability," said Deputy Marshal Willetts, "Graham county will probably offer not less than $1,000, the government and the state will add to it."

### Description of Hunted

The description of the killers is as follows:

John Powers, 23 years old, sandy complexion, smooth shaven, weight about 160 pounds, 5 feet, 8 inches tall, slender.

Tom Powers, 26 years old, dark complexion, weight about 160 pounds, 5 feet 8 inches tall; slender.

Tom Sisson, about 50 years old, very dark complexion, black mustache. Has rasping voice, about 5 feet, 11 inches tall, weighs about 180 pounds.

### Terrain of the Crime

The Galiuros are an exceedingly rugged range running Southeast and northwest, north of and aparallel to the Catalinas. Their imposing bulk rises before the traveler as his car descends into the San Pedro valley on the way to Mammoth or Reddington.

The Aravaipa canyon is a long and narrow cut in the hills to the West of the Galiuros, which reaches around back of them, opening into the Aravaipa valley on their farther side. Here are two small settlements, Aravaipa and Klondyke.

It is reported that the Powers brothers had sent word to the federal authorities at Safford that if they wanted them, they should come and get them.

**Report From Safford**

The Citizen is in receipt of a letter from Dr. H. A. Schell, the Tucson oculist, who happened to be at Safford Sunday, giving all the details known there at that time. He says that the Powers' brothers were located at a cabin in the mountains by the posse.

"Full particulars are not known," he continues, "but it has been learned that the father of the Powers boys was shot and possibly fatally wounded. The officers were evidently killed without warning while advancing towards the cabin. Word was brought in by Former Sheriff Haynes of Gila County, now a United States deputy marshal.

"The murderers made threats to kill any officers attempting to arrest them.

"Considerable excitement prevails at Safford.

"I heard that Sissons was out of the penitentiary on parole by Hunt, but afterwards heard that he had served his time for horse stealing and the one on parole is a different party."

J. A. McMurren of Tucson, a relative of "Maricopa Slim" Powers, special agent of the Southern Pacific, who was killed by tramps some years ago, denied reports that the Powers brothers are related to "Maricopa Slim", although the names are the same.

The Powers brothers are described in a circular calling for their arrest, issued by United States Marshall Dillon as the brothers of "Maricopa Slim".

**Account of Shooting**

According to Haynes' account, he planned to arrest the younger Powers on the draft evasion charge the Graham county officers volunteering to accompanying him that they might investigate the death of a sister of the alleged draft evaders, who died about six weeks ago, under circumstances which Haynes said the other officers considered suspicious. Haynes said that when the officers reached the Powers cabin in the mountains, after a long horseback and automobile trip from Safford, Arizona, they found Powers, his two sons and Sisson there. Haynes said they opened fire on the officers at once, and that after many shots had been exchanged McBride, Kempton and Wootan were killed and he was wounded. He hurried back to Safford to secure aid.

While he was gone, Powers' sons and Sisson took the horses and arms of the slain officers and escaped into the mountains.

The elder Powers, father of the two men alleged to to be evading the draft is reported dangerously wounded in the right eye.

# TRAP SET FOR POWERS GANG IN SOUTHERN MOUNTAINS

**Sheriff Miles Withdraws Forces from the North to Watch Southern Passes of Rincon and Catalina Mountains; Slayers Are Believed to Have Crossed the San Pedro River and Are Heading for Mexican Line.**

Convinced from the direction of tracks near Redington, that the Powers-Sisnicn outlaws are heading for the Rincon gap, the sheriff's forces are making a "squeeze play" from the north and south sides of the divide Tuesday in the hope of catching the murderers in between.

The latest information concerning the movements of the outlaws was brought to Tucson Tuesday afternoon by a Gila county posse led by Deputy Sheriff Reinhart. They said that tracks of Sheriff McBride's mule ridden by one of the outlaws, had been found south of Redington.

The tracks showed that the animal had cast three shoes.

It is believed that the mule and horses have played out and that the outlaws will make an attempt to steal fresh mounts. Reinhart had three men with him. After dining, they left for Agua Caliente to reinforce the Pima posses in the Rincon pass.

**Country Covered With Brush**

The country in the Rincon pass is said to be covered with thickets of underbrush to such an extent that a rider cannot see more than 15 feet ahead of him in many places.

There are now three Pima posses in the field. Deputy Sheriff Burts is leading men on horseback from Agua Caliente. Deputy Sheriff Ezekiels, L. E. Smith, H. R. Batterton, and others have gone to Tom Mills' ranch. Sheriff Miles is patrolling Happy Valley. Chief of Police Bailey Deputy Sheriff Archer and Cattle Inspector McKinney are with the sheriff.

A sister of Mr. Wootan by the A sister of Mr. Wootan by the name of Mrs. Hattie Francis, who lives near Fort Lowell, and her husband, who is employed by the Midvale Farms Co., are planning to leave for Safford to attend the funeral.

The Gila posse stated that the outlaws had cut barbed wire fences in two places.

The United States marshals office received a report late this afternoon that tracks thought to have been made by the bandits' horses and mule had been seen near San Simon, which is near the New Mexican line. If this is true, it would dispose of the theory that the men are headed south through the Rincons and Catalinas. It is known that the Powers formerly lived on a ranch east of San Simon and they may be headed in that direction.

Acting on information that the Powers gang of murderers have crossed the San Pedro river and are heading for passes in the Rincon mountains, Sheriff Miles organized civilian posses in Tucson Tuesday to assist in watching the trails in this section of the country. The probable scene of the expected battle draws closer to Tucson.

The passes which are regarded as most strategic are the old road across the saddle of the range between the Catalina and Rincon mountains, issuing near Agua Caliente, the Rincon canyon issuing at Tom Mills' ranch and Happy Valley, on the east side of the Rincon divide, reached by way of the Benson road.

The San Pedro valley being now well guarded by sheriffs' forces from the northern counties, Sheriff Miles withdrew his men from the Redington district Monday night, after 24 hours in the field, in order to watch his own territory on the southern slope of the Catalinas and western slope of the Rincons.

**Hard to Find Men.**

The sheriff was having some difficulty Tuesday in finding men who could spare time for picketing that might last for several days. Among the first to volunteer were J. C. Daly captain of the Fort Lowell home guards, L. E. Smith, secretary of the Tucson Rifle Club and county assessor, and Harry R. Batterton, his deputy. D. M. Penny, the attorney offered his automobile.

The inquest into the death of the three victims of the outlaws, Sheriff McBride, and his two deputies Mark Kempton and T. Kane Wootan, and Jeff Powers, father of the Powers boys, was in progress at Safford Tuesday. U. S. Deputy Marshal Haynes, the only one of the four officers to come out of the fight alive is there, being the chief witness.

Jeff Powers who was shot near the heart, died at the scene of the fight

# POWERS BANDITS ARE HEADED FOR TUCSON

(Continued from page one.)

a few hours later. The bodies were taken to Safford by horseback.

**Tracks in River**

Tracks of the fugitives have been discovered near Redington where they crossed the river. From the direction taken it is deduced that they are heading for the southern ranges, the Catalinas or Rincons. Reports that the Powers boys and Sisson were seen at Mammoth are without foundation, so far as known. It is believed by the officers unlikely that they would go in that direction.

County Attorney Chambers of Graham county, told the Citizen over the telephone Tuesday morning of a report at Safford that the killers had been surrounded in the northwestern corner of Cochise county. The man hunt is taking place in the corners of four counties. Pima and Graham, which overlap by one township, and Pinal and Cochise which are separated by that distance. This region is in the San Pedro valley, isolated, and inhabited only by small ranchers, mostly Mexicans.

**Mormons Up in Arms.**

Forces of Mormon settlers from Graham county, aggregating over 150 men are in pursuit of the outlaws, spurred on by the fact that two of the slain officers were members of that faith.

It was learned at Tucson Tuesday that Sisson frequently came to the Bayless commissary at Redington to buy supplies, although the Powers boys were not often there. They preferred to use the Redington postoffice, although Klondyke was nearer.

In a letter mailed from Redington on Feb. 4, T. J. Powers asked quotations from the Smith Sporting Goods company on two calibres of rifle catridges. The Powers had recently sought to get supplies of food from Redington.

**Wootman's Relatives Here**

William Wootan, father of the slain officer and Bras Wootan, his brother called at Sheriff Miles' office Monday afternoon. They were on their way by automobile from their home on Freeman's ranch, near Florence, to Safford.

Deming Isaacson, owner of the Hot Springs ranch, mentioned in the dispatch from Safford, was in Tucson Monday, a guest at the home of Byrd Brooks and was endeavoring to get in telephonic communication with his place.

**Story of the Fight**

An authentic story of the encounter, telegraphed to The Citizen by the Graham county guardian of Safford, is as follows:

Sheriff R. F. McBride and deputies, Mark Kempton and T. Kane Wootan were killed Sunday morning about 6:30 o'clock in a battle with Tom Powers, John Powers, slackers, and Tom Sissons, horsethief at the home of the Powers boys in Rattlesnake canyon, Galluro mountains.

Deputy United States Marshal Frank Haynes of Globe, Sheriff McBride and deputies Kempton and Wootan left Safford Saturday afternoon at 4:00 o'clock for the Galluro mountains by way of Klondyke to get Tom Powers and John Powers. They reached the house near the mine of John Powers, father of the Powers boys, about daybreak Sunday morning and surrounded the house. Old man Powers came out of the house with a gun, and was ordered to hold up his hands, as one of his boys opened the door and started firing at the officers. Old man Powers dropped, shot through the right shoulder about 25 shots were fired by the officers.

McBride was shot by Tom Powers and Kane Wootan by John Powers. Kempton came around the house, kicked in the door and was shot by Tom Sisson, the horse thief. Sisson was sentenced to the penitentiary 6 years ago and was paroled by Governor Hunt. He lived with the Powers.

**Hayes Was Unhurt.**

Marshal Haynes escaped unhurt and came back to Klondyke for help. A posse from Klondyke started Sunday afternoon for the Powers camp to bring back the bodies. Haynes arrived at Safford at 4:00 o'clock Sunday with the news of the killing.

The murderers took the officers horses and ammunition and started for the border to go into Mexico. They stopped at Hot Springs, one mile from the Powers camp and told a man named Allen they had killed the officers, and that their father was shot and for him to go to the Powers home and take care of the mine.

Allen and a man named Murdock went to the house and found the old man. They took him into the tunnel of the mine and placed him on a cot. Allen then came to Safford, reaching here about 4:00 o'clock Sunday afternoon. Posses left Safford, Thatcher, Pima, Fairview, and Fort Thomas Sunday night heavily armed for the Galluro mountains to get the slackers and Sisson. A report Monday from Redington in the San Pedro valley states that the Powers boys and Sisson reached there Sunday night.

The Citizen is in receipt of a letter from Ben T. Power of Winkelman, Ariz., a brother of "Maricopa Slim," Power, stating that John and Tow Powers are not members of the family. The name is spelled differently, the outlaws having an "s"

[ABOVE ARTICLE DATED FEBRUARY 13, 1918 REPRINTED COURTESY OF TUCSON CITIZEN]

[Transcript of article from Tucson Citizen, February 13, 1918.]

## TRAP SET FOR POWERS GANG IN SOUTHERN MOUNTAINS

**Sheriff Miles Withdraws Forces from the North to Watch Southern Passes of Rincon and Catalina Mountains; Slayers Are Believed to Have Crossed the San Pedro River and Are Heading for Mexican Line.**

Convinced from the direction of tracks near Redington, that the Powers-Sission [sic] outlaws are heading for the Rincon gap, the sheriff's forces are making a "squeeze play" from the north and south sides of the divide Tuesday in the hope of catching the murderers in between.

The latest information concerning the movements of the outlaws was brought to Tucson Tuesday afternoon by a Gila county posse led by Deputy Sheriff Reinhart. They said that tracks of Sheriff McBride's mule ridden by one of the outlaws, had been found south of Redington.

The tracks showed that the animal had cast three shoes.

It is believed that the mule and horses have played out and that the outlaws will make an attempt to steal fresh mounts. Reinhart had three men with him. After dining, they left for Agua Caliente to reinforce the Pima posses in the Rincon pass.

### Country Covered With Brush

The country in the Rincon pass is said to be covered with thickets of underbrush to such an extent that a rider cannot see more than 15 feet ahead of him in many places.

There are now three Pima posses in the field. Deputy Sheriff Burts is leading men on horseback from Agua Caliente. Deputy Sheriff Ezekiels, L. E. Smith, H. R. Batterton, and others have gone to Tom Mills' ranch. Sheriff Miles is patrolling Happy Valley. Chief of Police Bailey, Deputy Sheriff Archer and Cattle Inspector McKinney are with the Sheriff.

A sister of Mr. Wootan by the name of Mrs. Hattie Francis, who lives near Fort Lowell, and her husband, who is employed by the Midvale Farms Co., are planning to leave for Safford to attend the funeral.

The Gila posse stated that the outlaws had cut barbed wire fences in two places.

The United States marshals' office received a report late this afternoon that tracks thought to have been made by the bandits' horses and mule had been seen near San Simon, which is near the New Mexican line. If this is true, it would dispose of the theory that the men are headed south through the Rincons and Catalinas. It is known that the Powers formerly lived on a ranch east of San Simon and they may be headed in that direction.

Acting on information that the Powers gang of murderers have crossed the San Pedro river and are heading for passes in the Rincon mountains, Sheriff Miles organized civilian posses in Tucson Tuesday to assist in watching the trails in this section of the country. The probable scene of the expected battle draws closer to Tucson.

The passes which are regarded as most strategic are the old road across the saddle of the range between the Catalina and Rincon mountains, issuing near Agua Caliente, the Rincon canyon issuing at Tom Mills' ranch and Happy Valley, on the east side of the Rincon divide, reached by way of the Benson road.

**Hard to Find Men**

The sheriff was having some difficulty Tuesday in finding men who could spare time for picketing that might last for several days. Among the first to volunteer were J. C. Daly, captain of the Fort Lowell home guards, L. E. Smith, secretary of the Tucson Rifle Club and county assessor, and Harry R. Batterton, his deputy. D. M. Penny, the attorney, offered his automobile.

The inquest into the death of the three victims of the outlaws, Sheriff McBride and his two deputies, Mark Kempton and Kane Wootan, and Jeff Powers, father of the Powers boys, was in progress at Safford Tuesday. U.S. Deputy Marshal Haynes, the only one of the four officers to come out of the fight alive, is there, being the chief witness.

Jeff Powers who was shot near the heart, died at the scene of the fight a few hours later. The bodies were taken to Safford by horseback.

**Tracks in River**

Tracks of the fugitives have been discovered near Redington where they crossed the river. From the direction taken it is deduced that they are heading for the southern ranges, the Catalinas or Rincons. Reports that the Powers boys and Sisson were seen at Mammoth are without foundation, so far as known. It is believed by the officers unlikely that they would go in that direction.

County Attorney Chambers of Graham county, told the Citizen over the telephone Tuesday morning of a report at Safford that the killers had been surrounded in the northwestern corner of Cochise county. The man hunt is taking place in the corners of four counties, Pima and Graham, which overlap by one township, and Pinal and Cochise, which are separated by that distance. This region is in the San Pedro valley, isolated, and inhabited only by small ranchers, mostly Mexcians.

**Mormons Up In Arms**

Forces of Mormon settlers from Graham county, aggregating over 150 men are in pursuit of the outlaws, spurred on by the fact that two of the slain officers were members of that faith.

It was learned at Tucson Tuesday that Sisson frequently came to the Bayless commissary at Redington to buy supplies, although the Powers boys were not often there. They preferred to use the Redington postoffice [sic], although Klondyke was nearer.

In a letter mailed from Redington on Feb. 4, T. J. Powers asked quotations from the Smith Sporting Goods company on two calibres of rifle cartridges. The Powers had recently sought to get supplies of food from Redington.

**Wootman's [sic] Relatives Here**

William Wootan, father of the slain officer and Bras Wootan, his brother called at Sheriff Miles' office Monday afternoon. They were on their way by automobile from their home on Freeman's ranch, near Florence, to Safford.

Deming Isaacson, owner of the Hot Springs ranch, mentioned in the dispatch from Safford, was in Tucson Monday, a guest at the home of Byrd Brooks and was endeavoring to get in telephonic communication with his place.

**Story of the Fight**

An authentic story of the encounter, telegraphed to The Citizen by the Graham county guardian [sic] of Safford, is as follows:

Sheriff R. F. McBride and deputies, Mark Kempton and T. Kane Wootan were killed Sunday morning about 6:30 o'clock in a battle with Tom Powers, John Powers, slackers, and Tom Sissons, horsethief [sic] at the home of the Powers boys in Rattlesnake canyon, Galiuro mountains.

Deputy United States Marshal Frank Haynes of Globe, Sheriff McBride and deputies Kempton and Wootan left Safford Saturday afternoon at 4:00 o'clock for the Galiuro mountains by way of Klondyke to get Tom Powers and John Powers. They reached the house near the mine of John Powers, father of the Powers boys, about daybreak Sunday morning and surrounded the house. Old man Powers came out of the house with a gun, and was ordered to hold up his hands, as one of his boys opened the door and started shooting at the officers. Old man Powers dropped, shot through the right shoulder about 25 shots were fired by the officers.

McBride was shot by Tom Powers and Kane Wootan by John Powers. Kempton came around the house, kicked in the door and was shot by Tom Sisson, the horse thief. Sisson was sentenced to the penitentiary 5 years ago and was paroled by Governor Hunt. He lived with the Powers.

**Hayes [sic] Was Unhurt**

Marshal Haynes escaped unhurt and came back to Klondyke for help. A posse from Klondyke started Sunday afternoon for the Powers camp to bring back the bodies. Haynes arrived at Safford at 4:00 o'clock Sunday with the news of the killing.

The murderers took the officers [sic] horses and ammunition and started for the border to go into Mexico. They stopped at Hot Springs, one mile from the Powers camp and told a man named Allen they had killed the officers, and that their father was shot and for him to go to the Powers home and take care of the mine.

Allen and a man named Murdock went to the house and found the old man. They took him into the tunnel of the mine and placed him on a cot. Allen then came to Safford, reaching here about 4:00 o'clock Sunday afternoon. Posses left Safford, Thatcher, Pima, Fairview and Fort Thomas Sunday night heavily armed for the Galiuro mountains to get the slackers and Sisson. A report Monday from Redington in the San Pedro valley states that the Powers boys and Sisson reached there Sunday night.

The Citizen is in receipt of a letter from Ben T. Power of Winkelman, Ariz., a brother of "Maricopa Slim" Power, stating that John and Tow [sic] Powers are not members of the family. The name is spelled differently, the outlaws having an "s".

# POWERS GANG BELIEVED IN FOOTHILLS OF RINCONS

**Posses Are Searching Country in the Vicinity of Pool; Sheriff Wheeler and Men Are Proceeding Up Happy Valley; Outlaws Thought to Be Riding Same Mounts on Which They Fled**

**Sheriff Miles and posse left by automobile this afternoon for Poole, taking with them three bloodhounds, which will be placed on the trail of the Powers-Sission outlaws. It will take them about three hours to make the trip. The bandits are now believed to be in that vicinity. They went by the way of Benson.**

**BANDITS ELUDE POSSES**

The patient relentless trailing of the Powers-Sisson outlaws was continued Wednesday in the vicinity of Pool, a ranch settlement in the San Pedro valley by a posse from Pinal county and another from Cochise under Sheriff Wheeler. The gang is supposed to be in the foothills of the Rincons.

Sheriff Wheeler who was in Happy Valley on the eastern side of the Rincons Tuesday, is proceeding up the valley toward Pool. Deputy Sheriff Tom Burts, accompanied by Tom Wills, chairman of the board of supervisors of Pinal county, and others returned to Fuller Pass Wednesday. Burts and his men rode through the pass Tuesday as far as Vail's ranch and yet found no trace of travelers having been that way.

Tracks of the outlaws' animals were found in the vicinity of Pool about dark Tuesday evening. They have the same animals with which they started and are believed to be headed through Happy Valley.

**Trailed to Redington.**

Frank Wootan, a brother of the slain deputy and Hardenbrook, deputized by the acting sheriff of Graham county are in the posse that trailed the outlaws through the pass in the Galiuro mountains to within half a mile of the store at Redington.

At that point the Powers boys and Sisson apparently became scared for they doubled back a mile and a half and then detoured through the fields, cutting the barbed wire in four places with pliers to avoid ranch houses. Crossing the San Pedro they went up on the mesa and then came back to the river.

Wylie Morgan, an uncle of the Powers boys, who lives at Klondyke, was in Tucson Wednesday. He characterized Sisson as a pro-German and attributed to him and to the old man Powers, who was killed, the incitement of the young men to resist the draft.

**Sent Death Threats.**

The older Powers boy is said by Morgan to have tried to pass himself off as 32 years of age, above the draft age, but could not convince the authorities. Morgan warned Sheriff McBride that the boys would try to kill him if he went to their place. The boys are said to have sent word to Morgan to keep his mouth shut or they would kill him.

It was said Wednesday that the three outlaws are friends of Jack Garden, the well known prospector of the Catalina mountains.

Governor Hunt has offered a reward of $3,000 for their capture, and Graham county a reward of $1,000. A part of the rewards will be apid for the capture of any one of the men. It is expected that a reward will be offered by the federal government.

The rumor that the outlaws were in the vicinity of San Simon was run down by Deputy Sheriff Moore of Willcox and proven groundless. Constables Whitley and Olney followed the tracks of the animals until they were lost in a light snowfall. Inquiries at the old Powers ranch failed to develop any information as to their whereabouts.

[ABOVE ARTICLE DATED FEBRUARY 14, 1918 REPRINTED COURTESY OF TUCSON CITIZEN]

[Transcript of article from Tucson Citizen, February 14, 1918.]

# POWERS GANG BELIEVED IN FOOTHILLS OF RINCONS

**Posses Are Searching Country in the Vicnity of Pool; Sheriff Wheeler and Men Are Proceeding Up Happy Valley; Outlaws Thought to Be Riding Same Mounts on Which They Fled**

Sheriff Miles and posse left by automobile this afternoon for Poole [sic], taking with them three bloodhounds, which will be placed on the trail of the Powers-Sission [sic] outlaws. It will take them about three hours to make the trip. The bandits are now believed to be in that vicinity. They went by the way of Benson.

## BANDITS ELUDE POSSES

The patient relentless trailing of the Powers-Sisson outlaws was continued Wednesday in the vicinity of Pool, a ranch settlement in the San Pedro valley by a posse from Pinal county and another from Cochise under Sheriff Wheeler. The gang is supposed to be in the foothills of the Rincons.

Sheriff Wheeler, who was in Happy Valley on the eastern side of the Rincons Tuesday, is proceeding up the valley toward Pool. Deputy Sheriff Tom Burts, accompanied by Tom Wills, chairman of the board of supervisors of Pinal county, and others returned to Fuller Pass Wednesday. Burts and his men rode through the pass Tuesday as far as Vail's ranch and yet found no trace of travelers having been that way.

Tracks of the outlaws animals were found in the vicinity of Pool about dark Tuesday evening. They have the same animals with which they started and are believed to be headed through Happy Valley.

## Trailed to Redington

Frank Wootan, a brother of the slain deputy and Hardenbrook, deputized by the acting sheriff of Graham county are in the posse that trailed the outlaws through the pass in the Galiuro mountains to within half a mile of the store at Redington.

At that point the Powers boys and Sisson apparently became scared for they doubled back a mile and a half and then detoured through the fields, cutting the barbed wire in four places with pliers to avoid ranch houses. Crossing the San Pedro they went up on the mesa and then came back to the river.

Wylie Morgan, an uncle of the Powers boys, who lives at Klondyke, was in Tucson Wednesday. He characterized Sisson as a pro-German and attributed to him and to old man Powers, who was killed, the incitement of the young men to resist the draft.

**Sent Death Threats**

The older Powers boy was said by Morgan to have tried to pass himself off as 32 years of age, above the draft age, but could not convince the authorities. Morgan warned Sheriff McBride that the boys would try to kill him if he went to their place. The boys are said to have sent word to Morgan to keep his mouth shut or they would kill him.

It was said Wednesday that the three outlaws are friends of Jack Garden, the well known prospector of the Catalina mountains.

Governor Hunt has offered a reward of $3,000 for their capture, and Graham county a reward of $1,000. A part of the rewards wil be apid [sic] for the capture of any one of the men. It is expected that a reward will be offered by the federal government.

The rumor that the outlaws were in the vicnity of San Simon was run down by Deputy Sheriff Moore of Willcox and proven groundless. Constables Whitney and Olney followed the tracks of the animals until they were lost in a light snowfall. Inquiries at the old Powers ranch failed to develop any information as to their whereabouts.

# POWERS GANG ISLOCATEDIN CHIRICAHUA MOUNTAINS

**ews of Fight Between Outlaws nd Posses Expected at Any Hoer; Fugitives' Camping Place n the Busenback Ranch Near Dos Cabezas Discovered.**

The Powers-Sisson bandits have tained new mounts, having stolen ree horses from the Williams ch near Dos Cabezas last night, cording to a long distance teleone call today from Sheriff Welch Tombstone received by C. L. Johnes at Willcox. The animals cont of one bay and one sorrel horse d a brown mare. The bandit's cks have been found in Rucker nyon, in the Chiracahuas, and the ses are close behind, according to eriff Welch. It is believed the ndits will make a desperate atmpt to cross the Mexican border night and to prevent this patrols e been redoubled.

An unconfirmed report was cirlated today in Phoenix that a cowy had captured the bandits singleanded, but this is doubted at the eriff's office here.

**BATTLE BELIEVED NEAR**

The hunt for the Powers-Sisson laws now centers in the Chiricaa and Dos Cabezas mountains. The l and military forces in the field e been largely augmented in or to cover every conceivable avenue escape. A battle is expected at y hour.

Sheriff Miles of Pima county is tak-

Sheriff Miles of Pima county is tak-
g an active part in the hunt and is ing in the field. He and his ses have not returned to Tucn for several days. Deputy Sheriff rs of Tucson is out with a posse, t his location is not known.

Sheriff Henry Hall of Pinal coun- and former Sheriff Jeff Adams of ricopa county passed through Tucn Thursday evening en route from rence to the scene of the hunt. eriff Hall has recently returned m Rochester, Minn., where he unwent an operation by the Mayo others. In spite of his weakness m this stage of illness, he could

(Continued on page three, column six.)

# POWERS GANG STEAL THREE MORE HORSES

(Continued from Page One)

not longer restrain his impatience to take part in the chase since it was moving into the Chiricahuas, where he formerly ranched.

Jeff Adams, a well known and successful fighter of outlaws, also declared that he couldn't keep out of the hunt any longer and so he went over to Florence from his home at Phoenix. They gathered four or five men and started for the front.

Traveling only at night and camping during the day in the roughest canyons, the outlaws were camped Thursday in the pasture of the Busenback ranch, one mile from Dos Cabezas. The pasture is guarded by over a hundred possemen.

First reports of the sighting of the outlaws were given to possemen by a ranch woman, who said that the outlaws appeared at her house near Pearce, Arizona, and asked for food for themselves. She gave them some bread but their horses were unfed and had no water.

[Above article dated February 15, 1918 reprinted courtesy of Tucson Citizen]

[Transcript of article from Tucson Citizen, February 15, 1918.]

## POWERS GANG IS LOCATED IN CHIRICAHUA MOUNTAINS

News of Fight Between Outlaws and Posses Expected at Any Hour; Fugitives' Camping Place on the Busenback Ranch Near Dos Cabezas Discovered.

The Powers-Sisson bandits have obtained new mounts, having stolen two horses from the Williams ranch near Dos Cabezas last night, according to a long distance telephone call today from Sheriff Welch of Tombstone received by C. L. Johnson at Willcox. The animals consist of one bay and one sorrel horse and a brown mare. The bandits [indecipherable] have been found in Rucker canyon, in the Chiricahuas, and the posses are close behind, according to Sheriff Welch. It is believed the bandits will make a desperate attempt to cross the Mexican border at night and to prevent this patrols have been redoubled.

An unconfirmed report was circulated today in Phoenix that a cowboy had captured the bandits singlehanded, but this is doubted at the sheriff's office here.

### BATTLE BELIEVED NEAR

The hunt for the Powers-Sisson outlaws now centers in the Chiricahua and Dos Cabezas mountains. The civilian and military forces in the field have been largely augmented in order to cover every conceivable avenue of escape. A battle is expected at any hour.

Sheriff Miles of Pima county is taking an active part in the hunt and is camping in the field. He and his deputies have not reurned to Tucson for several days. Deputy Sheriff Burts of Tucson is out with a posse, but his location is not known.

Sheriff Henry Hall of Pinal county and former Sheriff Jeff Adams of Maricopa county passed through Tucson Thursday evening en route from Florence to the scene of the hunt. Sheriff

Hall has recently returned from Rochester, Minn., where he underwent an operation by the Mayo brothers. In spite of his weakness from this siege of illness, he could not longer restrain his impatience to take part in the chase since it was moving into the Chricahuas, where he formerly ranched.

Jeff Adams, a well known and successful fighter of outlaws, also declared that he couldn't keep out of the hunt any longer and so he went over to Florence from his home at Phoenix. They gathered four or five men and started for the front.

Traveling only at night and camping during the day in the roughest canyons, the outlaws were camped Thursday in the pasture of the Busenback ranch, one mile from Dos Cabezas. The pasture is guarded by over a hundred possemen.

First reports of the sighting of the outlaws were given to possemen by a ranch woman, who said that the outlaws appeared at her house near Pearce, Arizona, and asked for food for themselves. She gave them some bread but their horses were unfed and had no water.

# POWERS GANG TRAIL FOUND NEAR DOS CABEZAS

**Fugitives Believed to Have Taken Refuge in Cochise Stronghold, Forty Miles from the Border; Deputies and Civilians Guarding All Passes in Hope of Trapping the Desperadoes.**

The Powers-Sisson outlaws were reported near Dos Cabezas, Thursday morning. The report came from Charles Busenbark, who rode to the Adams ranch, from which place the report was telephoned to the deputy sheriff at Benson.

Busenbark found his barbed wire fence cut and trailed the parties into the foothills of Dos Cabezas mountains. He said they were going east, apparently aiming to strike the border East of Douglas.

Dave Adams, a cattleman, ranching 14 miles east of Douglas reported in Benson that an army aeroplane is being used in the search, that it has been hovering over that part of the country for two days and last night was using a searchlight.

**Government Offers Reward.**

The government has offered a reward of $600 for information leading to the arrest of the outlaws, or $200 for information leading to the arrest of any one of them.

Efforts are being made by the federal authorities to find an aeroplane for use in the hunt.

In an opinion rendered by Attorney General Wiley Jones as to the legality of the state's offering a reward for the outlaws dead or alive, he says that the law does not contemplate the payment of the reward except upon conviction. Boards of supervisors, however, have authority to offer a reward, he continued, and such offer should state that it is for the "apprehension" of the men wanted. If necessary force used in such apprehension should unfortunately result in their death he is still of the opinion that the reward would be payable.

The total amount of rewards offered is now $4,600.

The United States Marshal's office at Tucson has a confirmation of the report that the men were in the vicinity of Dos Cabezas.

**U. S. CAVALRY ON GUARD.**

(Associated Press Leased Wire)

BISBEE, Ariz., Feb. 14.—Two troops of United States cavalry and

(Continued on page three, column one.)

(Continued from Page One)

posses from half a dozen Arizona counties were co-operating today in an effort to capture Thomas Powers, John Powers and Thomas Sisson sought for the murder of three Graham county officers in a gun fight in the Graham mountains last Sunday.

Posses of civil officers reinforced by cowboys, early today were standing guard over the outlets of Cochise Stronghold, 40 miles north of the Mexican border, where the fugitives are believed to have sought refuge last night when hard pressed by Sheriff Rye Miles of Pima county. Meanwhile the cavalrymen, ordered out from Douglas, were patrolling the border to prevent the fugitives escaping into Mexico in case they should elude the possemen.

**NEAR COCHISE STRONGHOLD**

The pursuit of the draft outlaws, John and Tom Powers, and Tom Sisson, is drawing near Cochise stronghold, south of Benson, according to information received from various points in Cochise county on Thursday. The army patrols have gotten into the hunt now, two companies having been sent by Commander Morgan of Fort Huachuca to patrol the border of Cochise and to take a more active hand in the hunt if deemed advisable.

Sheriff Wheeler is closing in from the north and Guy Walch at the head of another posse on the east and south.

Sheriff Miles of Pima is following the trail with the bloodhounds from the Cross X ranch, north of the railroad, where the outlaws left one horse and a mule and stole two horses. The outlaws are presumed to have crossed the Southern Pacific and to be headed south and east. They have not yet been sighted.

Guy Welch is a deputy sheriff from Tombstone, Cochise county. His objective is Gleeson, near the stronghold. He has taken charge of several posses in that vicinity and detailed them to watch the eastern and southern sides of the Dragoon mountains.

**Officers From Near-by Towns.**

Officers have gone from Johnson, Pearce, Cochise, and Willcox to assist.

Cochise Stronghold is one of the natural forts of the Southwest. It was made famous by the Apache chief, Cochise, who terrorized Southern Arizona in the '70's.

Deputy United States Marshal McDonald has formed a posse of officers and citizens at Douglas to assist in the pursuit.

The place where the horse and mule were found is two miles from Deep Wells, 20 miles East of San Pedro.

[ABOVE ARTICLE DATED FEBRUARY 15, 1918 REPRINTED COURTESY OF TUCSON CITIZEN]

[Transcript of article from Tucson Citizen, February 15, 1918.]

# POWERS GANG TRAIL FOUND NEAR DOS CABEZAS

**Fugitives Believed to Have Taken Refuge in Cochise Stronghold, Forty Miles from the Border; Deputies and Civilians Guarding All Passes in Hope of Trapping the Desperadoes.**

The Powers-Sisson outlaws were reported near Dos Cabezas, Thursday morning. The report came from Charles Busenbark, who rode to the Adams ranch, from which place the report was telephoned to the deputy sheriff at Benson.

Busenbark found his barbed wire fence cut and trailed the parties into the foothills of Dos Cabezas mountains. He said they were going east, apparently aiming to strike the border east of Douglas.

Dave Adams, a cattleman, ranching 14 miles east of Douglas reported in Benson that an army aeroplane is being used in the search, that it has been hovering over that part of the country for two days and last night was using a searchlight.

## Government Offers Reward

The government has offered a reward of $600 for information leading to the arrest of the outlaws, or $200 for information leading to the arrest of any one of them.

Efforts are being made by the federal authorities to find an aeroplane for use in the hunt.

In an opinion rendered by Attorney General Wiley Jones as to the legality of the state's offering a reward for the outlaws dead or alive, he said that the law does not contemplate the payment of the reward except upon conviction. Boards of supervisors, however, have authority to offer a reward, he continued, and such offer should state that it is for the apprehension of the men wanted. If necessary force used in such apprehension should unfortunately result in their death he is still of the opinion that the reward would be payable.

The total amount of rewards offered is now $4,600.

The United States Marshal's office at Tucson has a confirmation of the report that the men were in the vicinity of Dos Cabezas.

**U.S. CAVALRY ON GUARD**
**(Associated Press Leased Wire)**

BISBEE, Ariz., Feb. 14.—Two troops of United States cavalry and posses from half a dozen Arizona counties were cooperating today in an effort to capture Thomas Powers, John Powers and Thomas Sisson sought for the murder of three Graham county officers in a gun fight in the Graham mountains last Sunday.

Posses of civil officers reinforced by cowboys, early today were standing guard over the outlets of Cochise Stronghold, 40 miles north of the Mexican border, where the fugitives are believed to have sought refuge last night when hard pressed by Sheriff Rye Miles of Pima county. Meanwhile the cavalrymen, ordered out from Douglas, were patrolling the border to prevent the fugitives escaping into Mexico in case they should elude the possemen.

**NEAR COCHISE STRONGHOLD**

The pursuit of the draft outlaws, John and Tom Powers, and Tom Sisson, is drawing near Cochise stronghold [sic], south of Benson, according to information received from various points in Cochise county on Thursday. The army patrols have gotten into the hunt now, two companies having been sent by Commander Morgan of Fort Huachuca to patrol the border of Cochise and to take a more active hand in the hunt if deemed advisable.

Sheriff Wheeler is closing in from the north and Guy Walch [sic] at the head of another posse on the east and south.

Sheriff Miles of Pima is following the trail with the bloodhounds from the Cross X ranch, north of the railroad, where the outlaws left one horse and a mule and stole two horses. The outlaws are presumed to have crossed the Southern Pacific and to be headed south and east. They have not yet been sighted.

Guy Welch is a deputy sheriff from Tombstone, Cochise county. His objective is Gleeson, near the stronghold. He has taken charge of several posses in that vicinity and detailed them to watch the eastern and southern side of the Dragoon mountains.

### Officers from Nearby Towns

Officers have gone from Johnson, Pearce, Cochise, and Willcox to assist.

Cochise Stronghold is one of the natural forts of the Southwest. It was made famous by the Apache chief, Cochise, who terrorized Southern Arizona in the '70's.

Deputy United States Marshal McDonald has formed a posse of officers and citizens at Douglas to assist in the pursuit.

The place where the horse and mule were found is two miles from Deep Wells, 20 miles East of San Pedro.

# BLOODHOUNDS ARE SENT TO HUNT DOWN BANDITS

**Chase for Powers-Sisson Gang Turned into State of Siege in the Chiricahua Mountains; Sheriff Wheeler Calls for More Volunteers; Pima County Dogs Are Rushed to Douglas.**

(Special to The Citizen)

BOWIE, Ariz.,—The Powers-Sisson bandits are afoot and surrounded in Rock Creek in the Chiricahuas. More than 500 possemen and officers have them surrounded and 50 more have just arrived from Globe and Miami. Local volunteers are leaving Bowie now for the scene of the final roundup. Large parties are headed that way from Clifton, Morenci, Silver City, and Lordsburg.

Wagon loads of provisions are being rushed to feed the possemen from Bowie and Douglas. Sheriff Wheeler of Bisbee is in personal charge and determined to catch the outlaws alive is possible.

STATE OF SIEGE

The outlaws who murdered Sheriff McBride and his deputies last Sunday morning in Rattlesnake canyon in the Galiuro mountains are, six days later, in a state of siege in the Chiricahua mountains in southeastern Cochise county, about 150 miles from the scene of the crime. Their presumptive location is in Rucker and Rock canyons according to the latest information from telephone points and Sheriff Wheeler, who is at home in Cochise and therefore in command of the civil forces is strengthening the cordon of men which has been thrown around their hiding place.

A call for as many volunteers as possible was issued by Sheriff Wheeler Saturday morning, according to a report reaching Benson. The recruits will be used to reinforce and relieve the small army of cowboys and ranchers now seeking to prevent the escape of the outlaws during the night.

**Soldiers Join In Hunt.**

It is estimated that there are between 500 and 1,000 men now engaged in the hunt. Two companies of United States soldiers from Douglas have arrived in the Chiricahuas and are under the general direction of the Cochise commander.

When the picket lines have been securely established Wheeler and a small band of picked men will warily penetrate the rocky and forested wilderness of the Chiricahuas which they know so well and stalk their game foot by foot.

It is now known that it will be a fight to the death. Tom and John

(Continued on page three, column three.)

[ABOVE ARTICLE DATED FEBRUARY 16, 1918 REPRINTED COURTESY OF TUCSON CITIZEN]

# BANDITS ON FOOT ARE TRAPPED BY POSSES

(Continued from Page One)

Powers and Tom Sisson, famished, wearied and made desperate by the relentless pursuit of the hornets of the law which their crime stirred up, have shown every determination not to surrender alive. Wheeler and his right hand captain, Sheriff Miles of Pima county are the kind of officers who are determined to take them alive if it is possible to do so.

The chance that the infuriated cowboys and ranchers whom they have deputized for the chase, will endeavor to pull off a necktie party also has been considered, so that there is a chance that there may be two fights, one to capture the men and another to protect them until they can be landed in a secure jail.

**Outlaws Have Field Glasses.**

The eluding of the posses that have attempted to head the outlaws off as they made their way down the San Pedro valley toward the border is attributed partly to the fact that they are in possession of two pairs of field glasses taken from the bodies of the slain officers and so could detect posses by sighting them from peaks, if traveling during the day and by avoiding campfires at night.

The call for volunteers from Wheeler asks that they bring bedding and provisions. The campaign has rapidly developed into such proportions that the commissary has become one of the problems of the commander.

"Rocket," the English bloodhound in which Sheriff Miles places hope of locating the exact whereabouts of the outlaws and two two young dogs, were shipped to Douglas Friday night after having been returned to Tucson.

The outlaws are now dismounted and it is hoped to find a fresh trail. Their horses, saddled and bridled, were abandoned Friday in Rucker canyon because they were worn out. The outlaws then went on, taking Rock Creek canyon, which branches from Rucker's canyon.

Circulars bearing the photographs, descriptions and notice of rewards are being distributed not only throughout Ariona, but also New Mexico and California, so that if the men escape from the Chiricahuas, the chase will not end.

It is believed that if the outlaws finally cross the line into Mexico their troubles will just have commenced as the reward would be very attractive to some of the adventurers in that country.

**Sent Back With Dogs.**

Deputy Sheriff Archer and Ranger Bryan were sent back to Tucson Friday by Sheriff Miles with the dogs. The officers are needed to attend to other work here.

Rocket, the old dog, has a reputation as a killer. In California, where he was formerly worker, a Mexican fugitive was caught without a tree handy and his throat torn open by the fierce dog, before the officers could get to the spot.

If the dogs locate the fugitives, they will probably be shot, but the maxim is better a dog than a man sacrificed and the sound of the shots will be the signal for which the officers are waiting.

Deputy Sheriff Putnam of Deming, N. M., is directing the operation of posses along the state line.

Fifteen hundred dollars additional rewards have been offered by cattlemen of Graham and Cochise counties, it is reported, making a total of $6,100 now offered for their capture.

The Mexican border is 27 miles from Rucker's canyon. It is a steep walled gash in the mountains.

A carload of horses for use of possemen was detrained at Bowie Friday.

A report from Douglas late Saturday afternoon was that 700 cavalrymen are now in the field.

United States Marshal Dillon passed through Tucson Friday night on his way from Phoenix to Douglas to assist in the campaign, especially as concerned the participation of federal forces.

[Transcript of article from Tucson Citizen, February 16, 1918.]

## BLOODHOUNDS ARE SENT TO HUNT DOWN BANDITS

**Chase for Powers-Sisson Gang Turned into State of Siege in the Chiricahua Mountains; Sheriff Wheeler Calls for More Volunteers; Pima County Dogs Are Rushed to Douglas.**

**(Special to the Citizen)**

BOWIE, Ariz.,—The Powers-Sisson bandits are afoot and surrounded in Rock Creek in the Chiricahuas. More than 500 possemen and officers have them surrounded and 50 more have just arrived from Globe and Miami. Local volunteers are leaving Bowie now for the scene of the final roundup. Large parties are headed that way from Clifton, Morenci, Silver City, and Lordsburg.

Wagonloads of provisions are being rushed to feed the possemen from Bowie and Douglas. Sheriff Wheeler of Bisbee is in personal charge and determined to catch the outlaws alive if possible.

### STATE OF SIEGE

The outlaws who murdered Sheriff McBride and his deputies last Sunday morning in Rattlesnake canyon in the Galiuro mountains are, six days later, in a state of siege in the Chiricahua mountains in southeast Cochise county, about 150 miles from the scene of the crime. Their presumptive location is in Rucker and Rock canyons according to the latest information from telephone hints and Sheriff Wheeler, who is at home in Cochise and therefore in command is strengthening the cordon of men which has been thrown around their hiding place.

A call for as many volunteers as possible was issued by Sheriff Wheeler Saturday morning, according to a report reaching Benson. The recruits will be used to reinforce and relieve the small army of cowboys and ranchers now seeking to prevent the escape of the outlaws during the night.

**Soldiers Join in Hunt.**

It is estimated that there are between 500 and 1,000 men now engaged in the hunt. Two companies of United States soldiers from Douglas have arrived in the Chiricahuas and are under the general direction of the Cochise commander.

When the picket lines have been securely established Wheeler and a small band of picked men will warily penetrate the rocky and forested wilderness of the Chiricahuas which they know so well and stalk their [indecipherable] foot by foot.

It is now known that it will be a fight to the death. Tom and John Powers and Tom Sisson, famished, wearied and made desperate by the relentless pursuit of the hornets of the law which their crime stirred up, have shown every determination not to surrender alive. Wheeler and his right hand captain, Sheriff Miles of Pima county are the kind of officers who are determined to take them alive if it is possible to do so.

The chance that the infuriated cowboys and ranchers whom they have deputized for the chase, will endeavor to pull off a necktie party also has been considered, so that there is a chance that there may be two fights, one to capture the men and another to protect them until they can be landed in a secure jail.

**Outlaws Have Field Glasses.**

The eluding of the posses that have attempted to head the outlaws off as they made their way down the San Pedro valley toward the border is attributed partly to the fact that they are in possession of two pairs of field glasses taken from the bodies of the slain officers and so could detect posses by sighting them from peaks, if traveling during the day and by avoiding campfires at night.

The call for volunteers from Wheeler asks that they bring bedding and provisions. The campaign has rapidly developed into such proportions that the commissary has become one of the problems of the commander.

"Rocket," the English bloodhound in which Sheriff Miles places hope of locating the exact whereabouts of the outlaws and two two [sic] young dogs, were shipped to Douglas Friday night after having been returned to Tucson.

The outlaws are now dismounted and it is hoped to find a fresh trail. Their horses, saddled and bridled, were abandoned Friday in Rucker canyon because they were worn out. The outlaws then went on, taking Rock Creek canyon, which branches from Rucker's [sic] canyon.

Circulars bearing the photographs, descriptions and notice of rewards are being distributed not only throughout Arizona, but also New Mexico and California, so that if the men escape from the Chiricahuas, the chase will not end.

It is believed that if the outlaws finally cross the line into Mexico their troubles will just have commenced as the reward would be very attractive to some of the adventurers in that country.

**Sent Back With Dogs**

Deputy Sheriff Archer and Ranger Bryan were sent back to Tucson Friday by Sheriff Miles with the dogs. The officers are needed to attend to other work here.

Rocket, the old dog, has a reputation as a killer. In California, where he was formerly worker [sic], a Mexican fugitive was caught without a tree handy and his throat torn open by the fierce dog, before the officers could get to the spot.

If the dogs locate the fugitives, they will probably be shot, but the maxim is better a dog than a man sacrificed and the sound of the shots will be the signal for which the officers are waiting.

Deputy Sheriff Putnam of Deming, N. M., is directing the operation of posses along the state line.

Fifteen hundred dollars additional rewards have been offered by cattlemen of Graham and Cochise counties, it is reported, making a total of $6,100 now offered for their capture.

The Mexican border is 27 miles from Rucker's [sic] canyon. It is a steep walled gash in the mountains.

A carload of horses for use of possemen was detrained at Bowie Friday.

A report from Douglas late Saturday afternoon was that 700 cavalrymen are now in the field.

United States Marshal Dillon passed through Tucson Friday night on his way from Phoenix to Douglas to assist in the campaigns, especially as concerned the participation of federal forces.

# BANDITS SLIP THROUGH NET CAST BY ARMY OF HUNTERS

**Search Is Being Continued Today in Cave Creek Canyon; Outlaws Steal Fresh Mounts; Posses Get So Close Fugitives Are Forced to Leave Newly-Prepared Meal and Flee.**

With nearly 1,500 deputies and civilians in pursuit, backed by more than 700 cavalry men from the border posts, Tom and John Powers and Tom Sissons, charged with the killing of three Graham county officers a week ago, have again eluded arrest and may be across the line in Mexico by this time, according to word received from the scene of the search today.

It had been confidently expected by the officers that the outlaws would be captured yesterday morning and elaborate preparations had been taken to prevent the fugitives slipping through the cordon thrown around the Southern Chiracahua mountains, where the hunted men had been forced to abandon their horses.

The $6,300 reward hung up for the capture of the gang dead or alive, seemed almost within the grasp of the officers, when word

alive, seemed almost within the grasp of the officers, when word

was received that the bandits' exhausted mounts had been found. The bloodhounds from Pima county were rushed to Rucker canyon and put on the trial, while hundreds of automobile parties from Douglas, Bisbee and New Mexico points hurried to the scene of the hunt yesterday determined to be in "at the death."

**Posses Are Disappointed.**

But the amateur and professional bandit-hunters were alike doomed to disappointment. Late last night word came from the scene of the hunt that the men had succeeded in stealing fresh mounts from a ranch near Cave Creek canyon and the chase was on again.

Sheriff Wheeler of Bisbee is acting as field marshal of the greatest man-hunt ever staged in the Southwest, and he is using all the wiles of his long experience along the border in an effort to trap the outlaws, but without avail.

Not a glimpse has been caught of the men since they fled from the Caliuro mountains a week ago Sunday after killing Sheriff McBride and his two deputies.

**Forced to Leave Meat.**

The posses yesterday apparently came closer to the outlaws than at any time since the chase began when they came upon an apparent camping place, which the bandits had fled, leaving freshly prepared food behind them. The indications were that the three men are still traveling together and have not separated as was for a time thought probable.

The search today is continuing in Cave Creek canyon, another of the defiles of the Chiracahua mountains. Meantime, heavy patrols are watching the border.

[ABOVE ARTICLE DATED FEBRUARY 18, 1918 REPRINTED COURTESY OF TUCSON CITIZEN]

[Transcript of article from Tucson Citizen, February 18, 1918.]

## BANDITS SLIP THROUGH NET CAST BY ARMY OF HUNTERS

**Search Is Being Continued Today in Cave Creek Canyon; Outlaws Steal Fresh Mounts; Posses Get So Close Fugitives Are Forced to Leave Newly-Prepared Meal and Flee.**

With nearly 1,500 deputies and civilians in pursuit, backed by more than 700 cavalry men from the border posts, Tom and John Powers and Tom Sissons [sic], charged with the killing of three Graham county officers a week ago, have again eluded arrest and may be across the line in Mexico by this time, according to word received from the scene of the search today.

It had been confidentially expected by the officers that the outlaws would be captured yesterday morning and elaborate preparations had been taken to prevent the fugitives slipping through the cordon thrown around the Southern Chiricahua mountains, where the hunted men had been forced to abandon their horses.

The $6,300 reward hung up for the capture of the gang dead or alive, seemed almost within the grasp of the officers, when word was received that the bandits' exhausted mounts had been found. The bloodhounds from Pima county were rushed to Rucker canyon and put on the trial [sic], while hundreds of automobile parties from Douglas, Bisbee and New Mexico points hurried to the scene of the hunt yesterday to be in "at the death."

### Posses Are Disappointed.

But the amateur and professional bandit-hunters were alike doomed to disappointment. Late last night word came from the scene of the hunt that the men had succeeded in stealing fresh mounts from a ranch near Cave Creek canyon and the chase was on again.

Sheriff Wheeler of Bisbee is acting as field marshal of the greatest man-hunt ever staged in the Southwest, and he is using all the wiles of his long experience along the border in an effort to trap the outlaws, but without avail.

Not a glimpse has been caught of the men since they fled from the Galiuro mountains a week ago Sunday after killing Sheriff McBride and his two deputies.

**Forced to Leave Meat.**

The posses yesterday apparently came closer to the outlaws than at any other time since the chase began when they came upon an apparent camping place, which the bandits had fled, leaving feshly prepared food behind them. The indications were that the three men are still traveling together and have not separated as was for a time thought probable.

The search today is continuing in Cave Creek canyon, another of the defiles of the Chiricahua mountains. Meanwhile, heavy patrols are watching the border.

# POWERS-SISSONS GANG TRAILED BY POSSES IN FRESH SNOW

**Hope of Capturing Bandits Is Increased by Storm in the Chiricahua Mountains**

(BY JOHN HALL)

GLEESON, Ariz., Feb. 19.—A fall of snow following on the heels of a gale in Sulphuer Springs valley Monday, this morning brought promise of a fresh opportunity to "cut" the signs of the Sisson and Powers outlaws, who are still somewhere in the Chiricauhas. The snow fall, while light here is believed to be heavy in the Chiricahua, 22 miles east of here.

The whereabouts of the outlaws was absolutely unknown yesterday afternoon at the headquarters of the hunt near Rucker canyon, when the correspondent there, Deputy Sheriff Burts, of Tucson, Supervisor Tom Wills, of Florence and others camped at Webster's ranch, two miles north of Rucker and five miles south of Rock canyon, waiting for word from Sheriff Wheeler, and his trackers back in the mountains.

The last heard of Wheeler, he was driving the outlaws southward. Acting on this hope, Burts with the bloodhounds struck into the mountains from Webster's on the chance of cutting a fresh trail.

**Penetrated Mountains On Foot.**

The party went by auto to the head of Webster's canyon, then penetrated the mountains on foot. Wheeler has with him two expert trackers, "Bill" Burleson, of Cochise, formerly engaged in trailing runaway boys from the Fort Grant Reform school, and Lee Adams, of Dragoon. Reliance is placed on two Indian trailers from San Carlos, expected this afternoon.

The outlaws recently threw the officers off the trail by tying gunny sacks over their shoes. The sacks were discarded when worn out by the rocks. The dogs have not had a fair chance.

A "bone head" was pulled in allowing "Bill" Wootan, father of one of the slain officers, to use the saddles abandoned by the outlaws before Burts brought the dogs from Douglas, so destroying the scent. When the dogs arrived Sheriff Wheeler said the trail was too old and dogs were not given a chance.

**Camp At Stafford's Ranch.**

Two hundred civilians and 400 soldiers camped at Stafford's ranch, Rock Canyon, Sunday night. Troops B, C, L and K of the first cavalry from Douglas moved from Stafford's to Ruckers Monday, camped five miles up and sent patrols into the mountains. The officers on the west of the range are out of touch with the posses on the east side.

Four troops of the tenth cavalry, negroes from Natches, N. M., are detailed to watch the east side of the range. The campaign is being conducted by a ranchers' telephone line running along the west side of the mountains. This line reaches Dos Cabezas, but is not connected with a private line from Dos Cabezas, to Willcox.

**Slaughter at Cave Creek.**

Sheriff Slaughter, of Greenlee, has arrived at Cave Creek, with a large posse. Corrigan's bloodhounds, from Grant county N. M., are held at Cottonwood Canyon, north of Rock canyon, waiting instructions.

Eight hundred soldiers and it is estimated 1,000 civilians now are engaged in the hunt. The First Cavalry is supplied with hay and supplies by four army motor trucks making a daily round trip from Douglas, 40 miles south.

This man hunt is on a greater scale than any ever before conducted in Arizona, said Supervisor Wills yesterday. The officers believe the law still has a chance to get the quarry but freely express the opinion that this chase has been bungled and that Wheeler has deterred the northern deputies from entering the mountains. All officers are wearing white rag bands on their hats to prevent mistaking them for outlaws.

Hundreds of autos are being used in the operations. All civilians are heavily armed, with rifles and revolvers and wrapped with cartridge belts.

The trailing from the cross X ranch to Busenbark's ranch, near Dos Cabezas, was done by Deputy Sheriff Bright of Johnson, Harry Kennedy of Bonita, and Dave Felshaw of Pima. Wheeler took up the trail from Busenbarks, into the mountains.

**Near Jial At Johnson.**

The outlaws passed within 20 feet of the jail at Johnson, said Bright. They camped one day in Cochise stronghold and passed within three-quarters of a mile of Renaud's powder mill at Pearce. They camped at Mud Spring and went from there to Busenbarks, and entered the moun-

(Continued on page three, column five.)

## SNOW AIDS HUNT FOR THE POWERS OUTLAWS

(Continued from page one)

tains between Rock and Turkey creeks.

Al Kuntz prospector and trapper from Klondyke who packed out the bodies of the slain officers, says old man Powers told the murderer's nearest neighbors while dying, that Kave Wooten shot him, while he had his hands up. Haynes' story is that Powers lowered his arms and Wootan, thinking he reached for his gun, fired. The story that Powers came out with a gun told by Haynes, was denied by Powers on his death bed.

Kountz is now cooking for the posse, at Webster's ranch. The body of old man Powers was left unburied for several days by incensed Klondykers and was finally buried at the mine by Kuntz upon instructions of the county attorney. Tom Powers fired through a pane of glass, killing Wootan. The bodies were found close to the corners of the house, where they fell.

[Above article dated February 20, 1918 reprinted courtesy of Tucson Citizen]

[Transcript of article from Tucson Citizen, February 20, 1918.]

## POWERS-SISSONS GANG TRAILED BY POSSES IN FRESH SNOW

### Hope of Capturing Bandits Is Increased by Storm in the Chiricahua Mountains

(BY JOHN HALL)

GLEESON, Ariz., Feb. 19—A fall of snow following on the heels of a gale in Sulphur Springs valley Monday, this morning brought promise of a fresh opportunity to "cut" the signs of the Sisson and Powers outlaws, who are still somewhere in the Chiricahuas. The snow fall, while light here is believed to be heavy in the Chiricahuas, 22 miles east of here.

The whereabouts of the outlaws was absolutely unknown yesterday afternoon at the headquarters of the hunt near Rucker canyon, when the correspondent there, Deputy Sheriff Burts, of Tucson, Supervisor Tom Wills, of Florence and others camped at Webster's ranch, two miles north of Rucker and five miles south of Rock canyon, waiting for word from Sheriff Wheeler, and his trackers back in the mountains.

The last heard of Wheeler, he was driving the outlaws southward. Acting on this hope, Burts with the bloodhounds struck into the mountains from Webster's on the chance of cutting a fresh trail.

**Penetrated Mountains On Foot.**

The party went by auto to the head of Webster's canyon, then penetrated the mountains on foot. Wheeler has with him two expert trackers, "Bub" Burleson, of Cochise, formerly engaged in trailing runaway boys from the Fort Grant Reform school and Lee Adams, of Dragoon. Reliance is placed on two Indian trailers from San Carlos, expected this afternoon.

The outlaws recently threw the officers off the trail by tying gunny sacks over their shoes. The sacks were discarded when worn out by the rocks. The dogs have not had a fair chance.

A "bone head" was pulled in allowing 'Bill' Wootan, father of one of the slain officers, to use the saddles abandoned by the outlaws before Burts brought the dogs from Douglas, so destroying the scent. When the dogs arrived Sheriff Wheeler said the trail was too old and the dogs were not given a chance.

**Camp at Stafford's Ranch.**

Two hundred civilians and 400 soldiers camped at Stafford's ranch, Rock canyon, Sunday night. Troops B. C. L. and K. of the first cavalry from Douglas moved from Stafford's to Ruckers [sic] Monday, camped five miles up and sent patrols into the mountains. The officers on the west of the range are out of touch with the posses on the east side.

Four troops of the tenth cavalry, negroes from Natchez, N. M., are detailed to watch the east side of the range. The campaign is being conducted by a ranchers' telephone line running along the west side of the mountains. This line reaches Dos Cabezas, but is not connected with a private line from Dos Cabezas, to Willcox.

**Slaughter at Cave Creek.**

Sheriff Slaughter, of Greenlee, has arrived at Cave Creek, with a large posse. Corrigan's bloodhounds, from Grant county, N. M., are held at Cottonwood canyon, north of Rock canyon, waiting instructions.

Eight hundred soldiers and it is estimated 1,000 civilians now are engaged in the hunt. The First Cavalry is supplied with hay and supplies by four army motor trucks making a daily round trip from Douglas, 40 miles south.

This man hunt is on a greater scale than any ever before conducted in Arizona, said Supervisor Wills yesterday. The officers believe the law still has a chance to get the quarry but freely express the opinion that this chase has been bungled and that Wheeler has deterred the northern deputies from entering the mountains. All officers are wearing white rag bands on their hats to prevent mistaking them for outlaws.

Hundreds of autos are being used in the operations. All civilians are heavily armed, with rifles and revolvers and wrapped with cartridge belts.

The trailing from the cross [sic] X ranch to Busenbark's ranch, near Dos Cabezas, was done by Deputy Sheriff Bright of Johnson, Harry Kennedy of Bonita, and Dave Felshaw of Pima. Wheeler took up the trail from Busenbarks [sic], into the mountains.

**Near Jial [sic] At Johnson.**

The outlaws passed within 20 feet of the jail at Johnson, said Bright. They camped one day in Cochise stronghold and passed within three-quarters of a mile of Renaud's powder mill at Pearce. The camped at Mud Spring and went from there to Busenbarks [sic], and entered the mountains between Rock and Turkey creeks.

Al Kuntz prospector and trapper from Klondyke who packed out the bodies of the slain officers, says old man Powers told the murderer's [sic] nearest neighbors while dying, that Kave Wooten [sic] shot him, while he had his hands up. Haynes' story is that Powers lowered his arms and Wootan, thinking he reached for his gun, fired. The story that Powers came out with a gun told by Haynes, was denied by Powers on his death bed.

Kountz is now cooking for the posse, at Webster's ranch. The body of old man Powers was left unburied for several days by incensed Klondykers and was finally buried at the mine by Kuntz [sic] upon instructions of the county attorney. Tom Powers fired through a pane of glass, killing Wootan. The bodies were found close to the corners of the house, where they fell.

# BANDITS HAVE SLIPPED FROM NET ADMITS WHEELER

**Sheriff in Charge of Man-Hunt for Powers Boys and Sissons Says They Evidently Have Escaped from the Chiricahua Mountains; Search Is Renewed with Apache As Base of Operations.**

(Associated Press Leased Wire)

DOUGLAS, Ariz., Feb. 20.—Sheriff Harry C. Wheeler of Cochise county, in charge of the searchers for Thomas and John Powers and Thomas Sisson, sought as slayers of three Graham county officers, was in Douglas early today, and said that the hunt for the fugitives in the Chiricahua mountains had been adandoned. He said he was convinced that the trio had slipped through the cordon of approximately 1,000 men who have been watching the mountain exits.

The headquarters of the trailers has been moved to Apache, Arizona, and this morning posses were sent north and west from there. It is believed that the hunted men are heading toward Mexico and both American and Mexican troops are closely guarding the international line.

Wheeler and Slaughter left for Apache.

Apache is a point on the El Paso & Southwestern railroad on the Southwestern edge of the Chiricahuas.

HUNT LEFT TO WHEELER

**HUNT LEFT TO WHEELER**

The hunt for the outlaws has now practically been left to Sheriff Wheeler of Cochise county, in which the Chiricahuas lie. All Pima county officers have returned from Webster's ranch near Rucker's canyon, where they were last stationed. A few of the Globe and Safford men are still sticking to the campaign.

Although the whereabouts of the outlaws in the mountains are now unknown, Sheriff Wheeler is reported from Douglas to have gone to Apache, a station on the El Paso and Southwestern railroad, with a plan for searching the southeastern flank of the range.

Three Indian trailers from San Carlos, who went into the mountains from Bowie are the next reliance of Wheeler for tracking the outlaws. All former trails have been destroyed by the heavy snowfall which blanketed the Chiricahuas Monday night, but this may be of assistance in striking a fresh trail should the outlaws move around or attempt to leave the range for another hiding place.

**Sisson The Leader.**

There can be no question that Tom Sisson, former soldier, who spent several years as a packer in the Chiricahuas for the army, is leader of the outlaws. He knows the range by heart, the trails and peaks, canyons and streams. A sea-

(Continued on page three, column five)

[Above article dated February 21, 1918 reprinted courtesy of Tucson Citizen]

# PIMA OFFICERS LEAVE PURSUIT TO WHEELER

(Continued from Page One)

soned camper and outdoors man, he will be able to make his band at home whether it snows or shines, winter or summer.

The keen eyes of the outlaws are believed by the deputy sheriffs to have been aided in avoiding the random posses by using the excellent field glasses taken from Sheriff McBride. Chance may result in their capture.

The Chiricahua mountains, a paradise of game, a domain for a Robin Hood with its running streams, its beautiful forests and its maze of canyons and peaks, has become an outlaws' forest as a result of the escape into its vast fastnesses of the outlaws. With the virtual acknowledgement by many officers who have engaged in the chase that the murderers of Sheriff McBride and his two deputies have made good their escape, for the present at least, the Chiricahuas become one place in Arizona where the law is defied.

**Ideal Retreat For Fugitives.**

This range of mountains makes an ideal retreat for fugitives from justice. It is a kingdom in extent, reaching for 40 miles north and south from the Dos Cabezas range in the north to the El Paso & Southwestern railroad in the south and its width is fully 25 miles. From the canyons in the foothills which have an altitude of 7,000 feet, great peaks rise to 10,000 feet in height.

The Chiricahuas were the scene last summer of a far reaching forest fire, which required a small army of men to subdue. Ten minutes after a hunter has disappeared from sight in its mazes, he would be lost to all pursuit except expert trackers. So long is its boundary that the range can scarcely be surrounded even by the hundreds of men engaged, so as to prevent the outlaws from leaving the retreat at their own time and at their own place, should they desire to reach Mexico or New Mexico or some other range where pursuit will not press so hard.

Chance has favored the game instead of the hunters so far.

**Civilians Losing Interest.**

How long the cavalrymen will be kept on the job is not known, but with the lack of definite information, the civilian posses are losing interest in the chase and are going home.

Deputy Sheriffs Burts and Ranger Cummins, who were needed for regular duties in Tucson, returned Tuesday afternoon with the Pima county dogs. Burts, Tom Willis, supervisor from Florence, Cummins, Leverton of Mammoth, Hamilton of Winkleman and Dr. Izard, veterinarian of Tucson, trainer of the dogs, spent Monday night high up in the Chiricahuas in the snowstorm. They had a campfire and some grub for company. They saw no signs of the hunted men. There were nine men in the party.

Rumors of fresh signs and dispatches telling of the movements of various posses will continue to come in from Bowie, Willcox and other points but news of the outlaws themselves will probably break in an unexpected quarter. Sheriff Wheeler appears determined to stick to the job, either in person or by deputy, until the men are caught or until he is as well satisfied as he can be that they have left the state. Since the outlaws were in Cochise county when last heard from it is considered up to Wheeler to finish the job.

**Globe Chief In Hunt.**

Robert Pinyan, chief of police of Globe, has joined the hunt, being located at a ranch 16 miles southeast of Bowie. United States Marshal Dillon went north from Douglas, Tuesday, to take a personal survey of the situation. He probably headed for Rucker's canyon where the First Cavalry camped.

Ranchers in the Sulphur Springs valley, which lies on the west side of the range, and in Rusker's canyon, showed their anxiety for the capture of the murderers, owing to their closeness to the scene.

[Transcript of article from Tucson Citizen, February 21, 1918.]

## BANDITS HAVE SLIPPED FROM NET ADMITS WHEELER

**Sheriff in Charge of Man-Hunt for Powers Boys and Sissons [sic] Says They Evidently Have Escaped from the Chiricahua Mountains; Search Is Renewed with Apache As Base of Operations.**

**(Associated Press Leased Wire)**

DOUGLAS, Ariz., Feb. 20.—Sheriff Harry C. Wheeler of Cochise county in charge of the searchers for Thomas and John Powers and Thomas Sisson, sought as slayers of three Graham county officers, was in Douglas early today, and said that the hunt for the fugitives in the Chiricahua mountains had been abandoned. He said he was convinced that the trio had slipped through the cordon of approximately 1,000 men who have been watching the mountain exits.

The headquarters of the trailers has been moved to Apache, Arizona, and the morning posses were sent north and west from there. It is believed that the hunted men are heading toward Mexico and both American and Mexican troops are closely guarding the international line.

Wheeler and Slaughter left for Apache.

Apache is a point on the El Paso & Southwestern railroad on the Southwestern edge of the Chiricahuas.

### HUNT LEFT TO WHEELER

The hunt for the outlaws has now practically been left to Sheriff Wheeler of Cochise county, in which the Chiricahuas lie. All Pima county officers have returned from Webster's ranch near Rucker's [sic] canyon, where they were last stationed. A few of the Globe and Safford men are still sticking to the campaign.

Although the whereabouts of the outlaws in the mountains are now unknown, Sheriff Wheeler is reported from Douglas to have gone to Apache, a station on the El Paso and Southwestern railroad, with a plan for searching the southeastern flank of the range.

Three Indian trailers from San Carlos, who went into the mountains from Bowie are the reliance for Wheeler for tracking the outlaws. All former trails have been destroyed by the heavy snowfall which blanketed the Chiricahuas Monday night, but this may be of assistance in striking a fresh trail should the outlaws move around or attempt to leave the range for another hiding place.

**Sisson The Leader.**

There can be no question that Tom Sisson, former soldier, who spent several years as a packer in the Chiricahuas for the army, is leader of the outlaws. He knows the range by heart, the trails and peaks, canyons and streams. A seasoned camper and outdoors man, he will be able to make his band at home whether it snows or shines, winter or summer.

The keen eyes of the outlaws are believed by the deputy sheriffs to have been aided in avoiding the random posses by using the excellent field glasses taken from Sheriff McBride. Chance may result in their capture.

The Chiricahua mountains, a paradise of game, a domain for a Robin Hood with its running streams, its beautiful forests and its maze of canyons and peaks, has become an outlaws' forest as a result of the escape into its vast fastnesses of the outlaws. With the virtual acknowledgement by many officers who have engaged in the chase that the murderers of Sheriff McBride and his two deputies have made good their escape, for the present at least, the Chiricahuas become one place in Arizona where the law is defied.

**Ideal Retreat For Fugitives.**

This range of mountains makes an ideal retreat for fugitives from justice. It is a kingdom in extent, reaching for 40 miles north and south from the Dos Cabezas range in the north to the El Paso & Southwestern railroad in the south and its width is fully 25 miles. From the canyons in the foothills, which have an altitude of 7,000 feet, great peaks rise to 10,000 feet in height.

The Chiricahuas were the scene last summer of a far reaching forest fire, which required a small army of men to subdue. Ten minutes after a hunter has disappeared from sight in its mazes, he would be lost to all pursuit except expert trackers. So long is its boundary that the range can scarecely be surrounded even by the hundreds of men engaged, so as to prevent the outlaws from leaving the retreat at their own time and at their own place [sic], should they desire to reach Mexico or New Mexico or some other range where pursuit will not press so hard.

Chance has favored the game instead of the hunters so far.

**Civilians Losing Interest.**

How long the cavalrymen will be kept on the job is not known, but with the lack of definite information, the civilian posses are losing interest in the chase and are going home.

Deputy Sheriffs Burts and Ranger Cummins, who were needed for regular duties in Tucson, returned Tuesday afternoon with the Pima county dogs. Burts, Tom Wills, supervisor from Florence, Cummins, Leverton of Mammoth, Hamilton of Winkleman and Dr. Izard, veterinarian of Tucson, trainer of the dogs, spent Monday night high up in the Chiricahuas in the snowstorm. They had a campfire and some grub for company. They saw no signs of the hunted men. There were nine men in the party.

Rumors of fresh signs and dispatches telling of the movements of various posses will continue to come in from Bowie,

Willcox and other points but news of the outlaws themselves will probably break in an unexpected quarter. Sheriff Wheeler appears determined to stick to the job, wether in person or by deputy, until the men are caught or until he is as well satisfied as he can be that they have left the state. Since the outlaws were in Cochise county when last heard from it is considered up to Wheeler to finish the job.

**Globe Chiem [sic] In Hunt.**

Robert Pinyan, chief of police of Globe, has joined the hunt, being located at a ranch 16 miles southeast of Bowie. United States Marshal Dillon went north from Douglas, Tuesday, to take a personal survey of the situation. He probably headed for Rucker's [sic] canyon where the First Cavalry camped.

Rancher in the Sulphur Springs valley, which lies on the west side of the range, and in Rucker's [sic] canyon, showed their anxiety for the capture of the murderers, owing to their closeness to the scene.

# 7

On the morning of the inquest, Jay Murdock and Ed Knothe arrived in Safford. Both had been called to testify, and both quickly sensed the ugly mood in town. Knothe's German accent gave him plenty of reason to worry, and Jay feared folks might think he was too friendly with slackers and murderers. Heads down and thrust forward in a peculiar syncopated movement, the two men walked slowly up the courthouse steps.

At two o'clock, a six-man coroner's jury of Safford's leading citizens viewed the bodies of McBride, Kempton, and Wootan. Arizona law required all bodies killed in a single, violent incident to be present at the same inquest, but Jeff Power's remains were conspicuously absent.

In the crowded courtroom, the scent of tung oil and dust hung in the stagnant air. Low, rumbling voices echoed throughout the somber chamber. When Justice of the Peace U. I. Paxton took his place at the bench, a hush fell over the room.

County Attorney Chambers rose from his chair and stood center stage. A vigorous, elfin man, he peered over his spectacles at the solemn faces in the jury box. This hearing was all Chambers' show. He had no adversary, no one to object to his questions or cross-examine his witnesses, but he had more than a few problems. No one claimed the posse had identified themselves as such to the men inside the cabin. Chambers had to prove that the men in the cabin could clearly see and identify the men outside. His major concern, however, was the fact he had no jurisdiction in the case. That detail had to be rectified if the ambitious County Attorney hoped to prosecute the Power brothers.

In a deep resonant voice, Chambers called Henry Allen to the stand. Young Henry seemed nervous and eager to please. In a thin, tremulous voice, he gave his name and explained his whereabouts on the morning of February 10.

Since none of his witnesses owned a watch, Chambers, with his next question, managed to push the time of the shootout forward to well after sunrise.

**Q: About 7:30 on Sunday, what did you hear?**
**A: We heard shootin'.**
**Q: Do you remember how many shots you heard?**
**A: No, sir, but I judge there was about twenty.**
**Q: Did you afterward see Tom Power, John Power, and Tom Sisson?**
**A: I seen them at Murdock's camp. They were on horseback. They had a brown horse, a sorrel horse, and a little sorrel mule. They rode right up with their guns drawn and told us what they done.**
**Q: Tell what each one—what did Sisson say?**
**A: I couldn't tell you just what he said. I said, "What's the matter, boys?" And John Power said, "The matter's layin'**

**over there in the yard. Frank McBride, his brother, and Kane Wootan is layin' over there."**

After the shootout, Tom and John, who could not identify Kempton's body, assumed he was McBride's brother, and that's what they told the men at Murdock's camp. At the mention of the bodies, the tension in the courtroom became palpable. Red began to creep out of Chambers' collar.

**Q: What did they say about the old man?**
**A: They says, "They killed father. He's just this side of the house; we'd like you fellas to take care of him, if he's not dead by this time."**
**Q: Did they say where they were going?**
**A: No, sir. I told them I'd get a horse and come on in. Tom Power looked straight at me and says, "What's that you say?" I says, "I'll go in and report what you fellas have done and try to get a doctor for your father." Tom says, "All right, go on."**

When Henry described the positions of the bodies in the yard, several women pulled out handkerchiefs to muffle their sobs.

In reference to a rumor that the bodies had been robbed, one tight-jawed juror asked Henry if he saw any evidence they had been disturbed when he first saw them.

**A: No, sir.**

The rumor that McBride had been shot in the head while he lay wounded and helpless on the ground prompted another question. The juror asked Henry if he heard a shot after the main shootout was over.

Again, Henry answered, "No, sir."

Chambers excused Henry and called J. A. Phillips, one of the vigilantes. The county attorney hoped to use Phillips to prove the Powers could identify by sight the men outside the cabin. The attorney took a who-could-best-see-whom approach to the problem.

Phillips described the location of the cabin as near the bottom of a deep canyon and facing east. The only door, he said, was canvas-covered and on the east side.

Actually, there were two canvas-covered doors on the east side.

Q: Did you notice how many windows were in the house?
A: I think four.
Q: Could a man on the outside be distinguished through the canvas door?
A: I think not. Well, now that is about sunup, about seven o'clock in the morning. These people being on the inside, any man's shadow could have easily been seen there.
Q: But from the outside you could probably not see a man on the inside?
A: No, you could not see a man on the inside. That would be a physical impossibility.

But what if the kerosene lamp on the kitchen table was lit? The opposite would be true. Of course, the supposition that a lawman can be identified as such by his shadow is ludicrous. Also, Phillips and Chambers failed to consider the weather conditions on the day in question. Under the overcast sky, the canyon was dark—so dark, Haynes would later admit, "You couldn't tell a man from a stump."

Phillips described the position of the bodies in detail. Chambers' voice quivered with anger.

Q: Did the bodies appear to have been disturbed after they had fallen?
A: Yes, they appeared to have been robbed. In fact, there was no money on them. Their purses, however, were left. Their overalls and pants were unbuttoned. I presume to give them access to their pockets.

The persistent juror asked, "Did you see anything near McBride's head to indicate he was shot after he fell?"

A: Yes, I seen that hole there.
Q: There was a hole there?
A: Yes, looked to me like you could stick two fingers in it.
Q: In the ground?
A: No, in his head.

A loud gasp swept through the courtroom.

Q: Not in the ground below his head?
A: Right in the pit of his ear.
Q: Any sign he'd been shot while he was layin' there?
A: No, I didn't notice anything that would indicate it on the ground. Why I think that happened after he'd fallen is for the reason that, at the corner where McBride was standing there was a chink in the log wall about a foot long at the height of a man's stomach, and in that room we found cartridge shells.

The grisly testimony drove several women from the courtroom, and some of the jurors looked mad enough to kill, but Phillips' testimony made no sense. How do chinks in the logs at the height of a man's stomach and cartridge shells add up to the conclusion that McBride was shot after he fell? It's possible Phillips himself may have been the source of the ugly rumor, as Chambers wisely declined to press his witness for an explanation.

The tension in the room was thick enough to slice. Chambers called Jay Murdock to the stand. Jay's father, John, still believing there would be an investigation into Jeff Power's death, had remained on the mountain to guard the body from predators.

**Q: Last Sunday morning did you hear shots from the direction of the Powers' place?**

**A: They were comin' so fast. Rifles, six-shooters, all at the same time like a machine gun—pop, pop, pop, just like that.**

**Q: After the shootout, did you talk to Tom Sisson, Tom Power, or John Power?**

**A: Yes, sir. They came up the trail on horseback. The first thing I said was, "It has happened." They says, "You're damn right it has happened. They shot father. He is wounded." They told me it was Kane Wootan, Frank McBride, and McBride's brother. They all began talkin'. Tom Power said his father stepped out, and they told him to throw up his hands, and he dropped the gun and threw up his hands, and Kane Wootan shot him. Then Kane turned around and shot through the window, and he filled Tom's eye full of glass—his left eye was all swollen up—and Tom says,**

"I took my rifle, punched the window out, and shot Kane Wootan."

Q: Did any of the others say anything then?

A: Yes, John Power says, "Well, I killed McBride."[4] I said, "What's the matter with your face? Have you been in a roughhouse?" He said it was splinters. He was bloody, and the blood was dried all over his face; it was just as bloody as it could be.

Leading the witness and pushing his luck, Chambers asked, "And Sisson said he shot one, too, did he?"

A: I didn't hear that part. Everyone was talkin' at once. I didn't hear anyone say Sisson killed the other man.

Q: How did you come to make the remark, "It has happened?"

A: McBride had given me a letter to take out to old man Power. I read the letter myself, and I was lookin' for McBride to come up. I told him I didn't believe the boys realized what it meant to do what they were doin'. McBride says, "Maybe I can explain and show the boys the mistake they are makin' by slackin' like they are." It was an awful fine letter explainin' to the boys what it meant to be a slacker and a traitor to their country, and so forth. He said, "It's not my duty as sheriff to arrest you boys, the government will take it up someday, and if you can get the boys to come in and give up," he said, "I think I can get them through without prosecution, and let them go on to the front and join the army." The next time I seen them, old man Power

[4] John believed Kempton was McBride's brother, and it is not clear to which McBride John referred.

says to me, he says, "Well, McBride says he isn't comin' up."

Q: Did they ever say what they would do if the officers came for them?

A: I had seen them—I suppose you remember they had a daughter livin' out there, Power's daughter, who died about a year or so ago. I saw them about two or three days after the funeral. I went to the house.

Chambers seemed surprised by Jay's testimony. After Tom and John failed to register for the draft, it was rumored they had left the cabin and were hiding from the law on the mountain or in Mexico.

Q: The boys were there?

A: Yes, sir, old man Power, Tom Sisson, John Power, and Tom Power, the four of them. The girl died, and she had been buried before I knew it. They sent for me, and I came over to see how this thing happened. They told me they hauled the body to Klondyke. No doubt the officers knew they were there, and I didn't feel safe sittin' there among those slackers. I made the remark, "Suppose McBride and his posse were to walk in here." Tom Power says, "Let him come. I'd just as soon die now as anytime, and while they're getting' me, I'll get some of them, maybe more than they get of us." Therefore, I always felt that at any minute somebody was goin' to come in there after them.

Q: Was that before you delivered the letter from McBride?

A: Yes, sir, maybe three weeks or a month before.

Q: Did you tell McBride about that?

A: Yes, sir. I told his Undersheriff, too. I told Mack like this, I said, "Mack, I've gotta live up there; I don't want to take

**sides, because if they get it in for me, they'd bump me off as quick as anybody else."**

**Q: Did they say anything more about what they would do if someone came after them?**

**A: Oh, yes, they told me they'd fight 'til they died and kill anyone who came for them.**

**Q: How did they get provisions up there?**

**A: I think they got them mostly from Safford and Klondyke. They had an automobile down in the Canyon, you know. They'd go down the mountain on horseback, then go in the car.**

**Q: Did you know of anybody carrying provisions from the other side of the mountain?**

**A: Not that I know of, no, sir. I don't think so.**

Frustrated by Murdock's response, Chambers quickly changed the subject.

**Q: When you found the bodies, did you recognize the dead men?**

**A: I knew two of them for sure, and the other one, Kempton, I couldn't tell if he was A. G. Welker or Mr. Kempton.**

**Q: What did you do with old man Power?**

**A: I cut his clothes off; I had him naked. I had to cut his clothes off before he died.**

Significantly, Chambers failed to ask a single question about the clothes Jeff wore. If Jeff Power had been fully dressed, that fact would support Chambers' claim that the Powers had lain in wait for the posse, whereas if he'd been wearing only his long johns, it would have supported the Powers testimony. But Chambers changed the subject.

Q: What did you do at the time you first went up there?

A: When father and I first got to him, he was unconscious. We straightened him out. He was layin' all crooked, and his eyes was glassy. He didn't know either one of us. I went up to get water and went back. The old man was conscious then. The first thing he said—he knew me—he said, "Give me a drink of water." I poured water down him. We put him in the mine tunnel. He didn't die until about four o'clock that evening.

Q: Did you stay with the body all night?

A: Yes, my father, myself, and Ed Knothe.

Q: Who is Knoche?

A: K-n-o-t-h-e. He's a German fellow, a little prospector.

Q: Had he been talking to these boys about staying out of the army?

A: I don't think so. No, he don't talk much.

Q: You never heard him express any opinion about the boys going to the army?

A: He told me he thought they were makin' a big mistake by slackin'.

Hearing his name mentioned, every nerve in Knothe's body seemed to leap and shudder. The implications behind Chambers' questions unnerved him. Pulling a red bandana from his hip pocket, he mopped his brow and sighed with relief when Chambers changed the subject.

Q: Now whereabouts was Jeff Power shot?

A: He was shot about there (indicating upper left breast). The bullet came out right between the shoulder blades.

Q: Could you tell the difference between where the bullet went in and came out?

A: Yes, sir, it was a bigger hole in the back than it was in front. He breathed through this bullet hole. I raised him up and the wind just fairly whistled through there. When I put my fingers over the hole, he could rest easier and breathe better. It bled awful bad.

Q: Did he say what happened before he was shot?

A: He said he didn't know there was anybody around. He says, "I stepped out in the yard, they hailed me and hollered, 'Throw up your hands.' I dropped my gun, throwed up my hands, says, 'We give up,' and Kane shot me." He repeatedly told me that all day.

Q: Did he say that was the first shot fired?

A: Yes, he said that was the first shot. He said, "I had my hands (up) beggin' them for peace."

Q: Did he seem to think he was going to die?

(The statement of a dying man is legal evidence, if the man knows he's dying.)

A: Yes, sir. I encouraged him all the time. He kept on sayin', "Jay, I'm goin' to die. I can't stand it." I said, "No, Mr. Power, you're good as a dozen dead men yet and shot too high to die. You'll get on all right just as quick as we get a doctor here."

If Chambers actually believed that Jay's encouraging words dispelled Jeff's premonition of death, he deluded himself. The trip to Safford and back took two days. Jeff Power had to know there would be no doctor before he bled to death.

The juror who was obsessed with McBride's head wound asked, "How long did the shooting last?"

A: About three minutes.

Juror: No more shooting?

**A: That's the last I heard.**

**Juror: Was there anything around McBride's head to indicate he was shot after he fell?**

**A: No, sir.**

Strangely enough, the questions Chambers chose *not* to ask were far more telling than the testimony he got from witnesses. Chambers failed to ask Jay what clothes Jeff Power was wearing, and he failed to ask Jay if he found evidence that Jeff's arms were raised when the bullet entered his body. Jay knew something about gunshot wounds, and he had plenty to say about the hole in Jeff's body, but Chambers didn't ask.

Over and over, Chambers set the time of the shootout at 7:30 when he questioned a witness. They were forced to guess at the time. Some estimates were just before sunrise, and some were just after, but no one mentioned the heavy cloud cover on the day in question, and Chambers didn't ask.

When Ed Knothe took the stand, Chambers questioned him about Sisson.

**Q: Did Sisson say anything?**

**A: Yes, he made some remark, but I can't recollect what he said.**

**Q: Do you remember if Sisson said he killed Frank McBride's brother?**

**A: I didn't hear that.**

Knothe had little to add to the testimony of Murdock and Allen. On the stand, sweat dripped from his forehead, and his face was colorless. When Chambers dismissed him, Knothe began to breathe again, and he looked like a man who'd trade his right arm for a stiff shot of rye.

Chambers, seeming deep in thought, paced back and forth in front of the bench. Then came the moment everyone was waiting for.

**"The court calls Deputy Marshal Frank Haynes."**

Every eye in the room was on the State's star witness as he strode down the aisle and took the oath. After Haynes identified himself and his job, Chambers asked:

**Q: Did you have occasion to come to Safford on the eighth or ninth of February for the purpose of getting some assistance to go into the Galiuro Mountains to arrest some slackers?**
**A: Yes, sir.**
**Q: What did you do when you came to Safford in reference to making the arrangements?**
**A: The arrangements was to be made by Mr. McBride. He was interested in court that day. That was Thursday. Friday we put in around here, and Saturday Mr. McBride, Mr. Kempton, Kane Wootan, and myself started over. I had authority from the United States Marshal to arrange for one posse. I believed he termed it in his letter "*a posse*," so that, according to the arrangements, was to be Mr. Kane Wootan.**
**Q: You had authority to get only one man on your responsibility?**
**A: Yes, sir. I had the warrants for the Power boys, and Mr. McBride had a warrant for old man Power and Sisson.**

A few words of perjured testimony, and Chambers' problems regarding jurisdiction disappeared as if by magic. According to Haynes, there were two posses at the Power cabin on that fateful day. The State warrants for Jeff and Sisson that Haynes mentioned separated McBride from the federal posse. Chambers could then claim jurisdiction over McBride's death. These mysterious warrants also explained Kempton's unauthorized presence at the cabin. No longer the odd man out, Kempton became McBride's posse. The warrants also changed Jeff Power's status. As a wanted man allegedly resisting arrest, his death at the hands of an old enemy, Wootan, could now be justified.

Only one problem remained—for what crime were Jeff and Sisson wanted? Haynes didn't say, Chambers didn't ask, and no one produced the alleged warrants in court.

Haynes went on to describe the posse's trip to Klondyke and their brief stay at Wootan's ranch. What he failed to mention was their stop at Habys'. Recounting the posse's winding descent down the hill behind the Power cabin, Haynes said, **"Just as Frank McBride and I came to the corner, I heard a voice I took to be Kane's holler, 'Throw up your hands.' I think he hollered twice after McBride and I come in sight. We passed the corner about four or five feet. This one man is the only man I seen, and McBride says, 'Boys, boys, boys,' just like that. Then I realized there was a man in the door; he had a Winchester."**

**Q: Did you see him?**

**A: Yes, he was in the doorway. This door hinged on the side toward McBride and I. Just when that opened, four or five shots came all in a bunch. The first shots came from the house. This old man fell.**[5]

[5] A picture of the cabin reportedly taken around the time of the shootout clearly shows the door hinged on the side where Wootan and Kempton stood.

Haynes' description of the scene seems deliberately vague. He said he saw two men. One in the doorway, and the location of the other, presumably Jeff Power, was anybody's guess. According to Jay Murdock's testimony, Jeff had just stepped into the doorway when somebody shouted, "Throw up your hands." Haynes didn't mention Jeff's gun. He said the first shots came from the house, and "This old man fell." According to this testimony, both sides agreed that the first shot hit Jeff Power, so the question of who fired first is easy to determine. Because the bullet entered Jeff's upper left chest and exited between his shoulder blades, it's a clear indication he faced the man who shot him. Unless he opened the door and backed out of the cabin, he was not shot by anyone inside the cabin.

Haynes continued: "Frank McBride and I hadn't fired yet. Then we both fired into this door."

Q: Into the door?

A: Yes, we both fired through the door at the man inside with the gun. I could see the gun and his two hands. The barrel of the gun was fast to the door shutter.

Q: Did you see him when he fired?

A: I could just see the opening in the door. This man fired at us. We fired again, and this man in the door—it looked to me like he turned his gun over and fired at Kane and Mart. He turned it over and back again. Kane and Mart went behind the corner of the house, and Frank and I took the other corner.

Q: You saw both Kane and Mart after the shooting began?

A: Yes, sir. These first shots came all in a bunch; then there had been four shots after that before they went behind the corner of the house. Well, McBride and I had arranged that, if anything tightened up there, to take the corners

> of the house. McBride motioned to me, and I says, "Shall I get to the other corner?" He said, "Yes." I see there is a window at that end of the house, but I couldn't see into the house, couldn't see inside at all. Well, I realized the noise had ceased on the other end of the house, and I asked Frank where Kane was, and he didn't answer me. Then I backed off a little further—there was some boards nailed up there on some posts—and I see McBride still on the corner. He fired two shots, it might have been three, then I didn't see McBride. Everything was as quiet as could be. In the meantime, somebody had fired a shot that I thought came out of the window at the end of the house where we was. It hit these boards. I turned around and fired to that window. McBride was still up then. I walked down by the side of these boards, and McBride was layin' on his back with his left leg stretched out and his left arm right down by the side of his leg and the other leg drawn up. I was afraid to go around the corner far enough to see his face.

The verbose Deputy Marshal held the audience in the palm of his hand. After all his exasperating minutiae about the boards, he finally got to the point, but forgot pertinent details. According to Haynes, he was never more than a few feet from McBride. During that time, three shots ripped through McBride's body before the head wound killed him. One shot went through his knee and crippled him. Two shots entered his torso, one of which exited through his back. The man was covered in blood. Haynes never mentioned McBride's wounds. Bullets tore through the sheriff's flesh and shattered bone. If he cried out, moaned, or even whimpered, Haynes seemed blissfully unaware. He described the position of the body, but never mentioned the blood or the gaping hole in the corpse. No one can say with certainty when Haynes fled from the cabin, but he didn't know his men were dead until he met Henry Allen in Klondyke. To believe Haynes actually witnessed the shootout requires a giant leap of faith.

The audience sat transfixed by Haynes' testimony. He left the courtroom, and Chambers called Doctor Platt to the stand. The doctor had been up all night. His rumpled faced sagged with fatigue and dark shadows circled his eyes.

Q: I will ask you first to take McBride's body and state what you found.
A: Frank McBride was shot in the right knee, right through the joint. Not directly through, but rising an inch and a half, and it came out about an inch higher on the opposite side. The shot below was two and a half inches below the upper one and didn't go through. A ball entered the right side of the ear and came out on the same angle, same level on the opposite side of the neck.

The courtroom was deathly silent. Chambers looked out on a sea of gray faces in the jury box.

Q: How many of the shots would have produced death?
A: Three
Q: Immediate death?
A: No, sir. He was probably living until he received the shot in the head.
Q: Now, in reference to Kempton, what did you observe?
A: Kempton received only one shot. It entered exactly in the center of the neck and carried away the first, second, and

**third vertebrae. Come out the side of the jaw and tore the whole side of the jaw out.**

Muffled sobs in the gallery broke the deadly silence.

**Q: Would the wound produce instant death?**
**A: Instant death, yes, sir.**
**Q: How many bullet wounds did you find on Wootan?**
**A: One: the ball entered to the left of the median line in the third vertebrae and exited an inch and a half from the sternum and tore out a hole as big as three silver dollars.**
**Q: Wootan was shot from the back?**
**A: Yes, sir.**
**Q: Did the wound produce instant death?**
**A: Instant death, yes, sir. He also had a bruise on his face that looked like the heel of a boot.**
**Q: Notice anything like that on McBride?**

Platt's mind remained fixed on Wootan's bruise. "It was undoubtedly produced before he died. It showed marks of injury before death."

Chambers tried to steer Platt back to McBride's wounds with only minor success.

**Q: Didn't notice anything like that on McBride?**
**A: No, I didn't notice any. Kane Wootan had another mark on the left hand which showed the heel injury, and the rest of the body was scarred in two or three places. I don't know whether that was produced before or after death. The one looked like it was produced before death, and the other looked like it might have been after.**

Platt's testimony evoked visions of shocking brutality, but the significance he seemed to attach to the injuries was not logical. Platt said Wootan died instantly. Since bruising cannot occur in the absence of blood pressure, the heel bruise was inflicted before or at the moment of death. If Platt was suggesting a man inside the cabin came out and kicked Wootan, an armed man, in the face while he was alive, how and when did it happen? As for the marks and scarring that occurred after death, the body had traveled face down on a packhorse through twenty-five miles of rough mountain terrain.

The jury, stunned by Platt's gruesome testimony, retired to consider their verdict. As expected, they found the officer had been killed in the lawful performance of duty by the men inside the cabin. Chambers claimed the Power brothers and Sisson had each killed an officer, and he planned to try each murder separately. Not one witness had incriminated Sisson, but Chambers filed first degree murder charges against him for the death of Sheriff McBride. On the complaint signed by Chambers, Tom and John Powers were named as Sisson's accomplices.

The stage was set for the trial of the decade, and a guilty verdict was a virtual certainty. Only one obstacle remained; they first had to catch the fugitives.

# 8

On the day after the inquest, the county mourned. *The Graham County Guardian* estimated 2,000 people attended combined services for McBride, Kempton, and Wootan. Black ribbons hung on the closed doors of every business on Main Street in Safford. Symbols of patriotism were everywhere. Layton Hall, decorated in red, white, and blue bunting, was filled to capacity with mourners. A large floral wreath stood by the lectern. The ribbon read, "Our Faithful Heros." Three flag-draped coffins rolled down the aisle. Three widows and nineteen fatherless children filled the first rows. Stationed at center front on a raised platform, dignitaries of church and state paid solemn tribute to the martyred officers.

Wootan and Kempton were laid to rest in neighboring graves in the Safford Cemetery. A pregnant Clara McBride accompanied her husband's body to the graveyard in Pima. McBride was buried with honor in his hometown. The epitaph on his headstone reads, "Faithful to duty until death."

The Graham County Board of Supervisors issued a written proclamation of sympathy to the grieving families.

On the same day, *The Tucson Citizen* reported that Jeff Power was buried near the mine tunnel by order of Graham County Attorney Chambers. There had been no autopsy nor inquest into his death, and there never would be.

According to newspaper accounts, a few days after the funeral, Charlie Power returned to the mountain to protect his family's mine from interlopers. He was too late. Laura Wootan, Sena Kempton, and Clara McBride had hired lawyer Spriggs. On behalf of the widows, Spriggs filed suit against Tom and John Power and the estate of Jeff Power for the wrongful death of the officers.

# 9

Several posses left the valley in search of Tom, John, and Sisson. In cars and on horseback, they raced across the desert and mountains to avenge the deaths of the officers.

The Graham County Board of Supervisors offered $1,000 for the capture of all three fugitives, or one-third of the amount for each. Governor Hunt, his narrow victory over Campbell still fresh in his mind, promised $1,000 for the capture and conviction of each "murderer." The total amount of rewards for the fugitives rose daily.

An editorial of *The Tucson Citizen* blamed the shootout on Hunt and the abolishment of capital punishment in Arizona. The editor chastised Graham County voters for their support of Hunt's "legal experiment." The Governor's liberal policies, the editor claimed, were responsible for Sisson's early parole, and he demanded the governor pay the rewards out of his own pocket.

According to Arizona law, rewards were paid upon conviction. It was illegal to offer rewards for suspects "dead or alive." However, the Attorney General announced the rewards were payable even if necessary force resulted in the death of a wanted man.

Impassioned reporters added to the hysteria. Wild rumors and unsubstantiated stories appeared in print daily. Reporters spoke of the "wanton massacre" of officers, and *The Tucson Citizen* claimed, "It is supposed the party of officers was ambushed, or the outlaws shot without warning." According to the article, Jeff Power was not dead, but wounded in the right eye, and the outlaw Sisson also suffered injury. Haynes' ever-changing accounts of the shootout didn't help matters, nor did Marshal Dillon's blunders. Dillon issued a rewards poster authorizing "all necessary force" in the capture of the fugitives. Tom and John Power, Dillon claimed, were the brothers of Maricopa Slim Power, a railroad agent murdered by tramps. Outraged relatives of Maricopa Slim quickly denied the story and demanded a retraction.

Meanwhile, the fugitives were badly in need of food, water, medical care, and fresh horses. Newspapers reported sightings of the men at widely separated locations. They were supposedly spotted in Redington on the west side of the Galiuros and in San Simon on the east end of the Pinaleño range at approximately the same time of day.

From the south, Sheriff Harry Wheeler of Cochise County moved into the San Pedro Valley near Redington. Sheriff Rye Miles from Tucson and a posse from Gila County closed in from the west and north. Believing they had the fugitives surrounded, Miles and Wheeler prepared for a bloody gun battle. In San Simon, Deputy Marshal Haynes searched for the fugitives from a safe distance.

During the long weeks of the search, Haynes, "the only living witness to the shootout," was busy obliging reporters with imaginative, inaccurate stories of the gunfight. County Prosecutor Chambers was no doubt suffering fist-pounding headaches resulting from his star witness's mindless ranting. Chambers had claimed that each fugitive killed one officer. He had formally charged Sisson with the murder of McBride. Now, according to Haynes' dramatic and widely published tale, Kempton had kicked in the back door, stormed the cabin, and Sisson had shot Kempton. Unfortunately, forensics proved Kempton was shot where he fell at the southeast corner of the cabin. Furthermore, the cabin had no back door, but Chambers couldn't call his star witness a liar.

Back in the San Pedro Valley, the fugitives slipped past the posse. Sheriff Miles withdrew forces in the north and staked out passes in the Rincon and Catalina Mountains. Wheeler, believing the trio had crossed the San Pedro River, also turned south.

One moonless night, the fugitives exchanged a tired sorrel and a tuckered mule for two fresh horses from the Cross X Ranch in Cochise County. Rye Miles moved in to track the men with bloodhounds. Deputies guarded the area around the Dragoon Mountains and the entrance to the famous Cochise Stronghold, a natural and nearly impenetrable fort the Apache Chieftain once used to hold off the American Army.

Troops stationed on Arizona's border with Mexico joined the largest manhunt in Arizona history. The army sent an airplane equipped with a floodlight to search the vast area, but no one sighted the slippery trio. Frustrated deputies declared German sympathizers were supplying the "Power Gang" with food and water.

Actually, the fugitives had little to eat or drink. It was not food and water that sustained them. It was the determination and stubborn pride indigenous to the Power family. Despite thirst, starvation, and agonizing pain, they refused to fall prey to a civilian posse behaving like a pack of wolves in a feeding frenzy. They intended either to surrender to the U.S. Army on the border, or cross into Mexico, and that, by God, is what they'd do.

Newspapers reported that Sisson, old army scout and "notorious horse thief," led the Power Gang over old Indian trails that no one else knew. Later, Tom laughed at the story. Every time they followed Sisson, Tom said, they got lost. Nevertheless, the story made sense to Cochise County Sheriff Wheeler. He brought in Indian scouts who found tracks in the snow on the Chiricahua Mountains east of Willcox. Weeks had passed since the search began, but Wheeler expected to capture the trio any time now, barring another snowstorm.

Under siege by the posse, the fugitives abandoned their starving horses on a grassy slope near a stream. Scrambling over rocky ridges with the soles of their shoes disintegrated, cold, tired, sick, and hungry, they continued to move south.

By the end of the week Wheeler had to admit the trio had slipped through his net again. The hunters, nearly as cold and miserable as the hunted, lost enthusiasm. The fugitives had been in Wheeler's territory from the beginning, but his moment of triumph had passed. By jealously guarding his jurisdiction, Wheeler had alienated other officers and undermined morale. Rye Miles and others went home, leaving the hunt to Wheeler.

One night, as the moon rose over the Chiricahuas, three men descended the mountain. Tom, John, and Sisson pushed southeast through the desert. Binding their torn shoes with rags, they slipped silently into Mexico. In the days that followed, they trudged over thirty miles into the Mexican desert, but found no water.

Parched and starving, they could barely stand. Tom suffered agony from his eye wound, but John was in worse shape. His eye was infected, and the pain was unbearable. Unable to survive without water, they turned back. The trio stumbled toward the border, dragging their feet and resting every few yards.

On March 8, nearly a month after the shootout, the exhausted men collapsed south of the border beneath a clump of mesquite. Eight miles to the north, a cavalry troop from Hachita, New Mexico discovered footprints leading south. Lieutenant Woolcott P. Hayes ordered his men to cut through the border fence and follow the trail.

Thirty years later, Brigadier General Woolcott P. Hayes described his illegal trek into Mexico in a letter to Graham County historian W. R. Ridgeway. The following is an excerpt from his letter.

"I was following the trail on horseback down a slight draw when I called to my corporal that I was still on the trail. When I turned my head to the front, I found myself looking down the three biggest gun barrels I had ever seen. The three men were crouched behind a thick clump of mesquite. I just fell off my horse and reached for my gun. The three of them could have killed me with ease, but did not shoot. They put their guns down and stood up. We saw what bad shape they were in. One of the boys had been shot in the eye. The whole socket was full of maggots. I gave them water and some dry crackers. Since our prisoners were not able to walk, we mounted them on our horses and started back to the border. I wanted to know why they had not resisted, and they said they would not shoot

# $4,000 REWARD

FOR CAPTURE OF

JOHN G. POWERS

## TOM POWERS

25 years old; dark complexion; smooth shaven; 5 feet 10 inches tall; slender; weight about 165 pounds. Dark hair, trifle curly. Is wearing bandage on forehead, evidently slightly wounded in fight with officers.

## JOHN POWERS

27 years old; dark complexion; 5 feet 9 inches tall; slender; weight about 165 pounds. Light hair, blue eyes.

TOM POWERS in Center

## TOM J. SISSON

Paroled convict, Arizona State Prison. 49 years old; 5 feet 9¼ inches tall; weight, 165 pounds; dark complexion; black hair; brown eyes; high forehead, deeply wrinkled. Rather stoop shouldered. Fairly good teeth. Lump on left side of head near the ear. Wears 7⅛ hat, and while in prison wore a heavy black beard.

These men killed Sheriff McBride, his undersheriff, M. R. Kempton, and Posseman Kane Wooten, while resisting arrest on a Federal warrant, charging them with evading the draft. The fight occurred at an abandoned mine near Klondyke, Graham County, Arizona, at 7:30 o'clock Sunday morning, February 10, 1918, and the murderers escaped, riding South on a sorrel horse, a brown horse and a sorrel mule. All animals carried full cowboy equipment. The County of Graham, Arizona, has offered a reward of One Thousand Dollars for the capture, dead or alive, of the three men, or one-third of this amount for either of them.

The Governor of Arizona has issued his proclamation offering a reward of One Thousand Dollars for each of these men subject to the provisions of the Arizona statutes, which authorize officers to use all proper necessary force to effect such apprehension.

The United States Government is expected to add materially to these offers of reward, having been requested by me to do so.

Advise me if any persons answering the description of these men are seen. Use the wires at my expense, Government rate.

Wire or write

## JOSEPH P. DILLON, U. S. Marshal

TUCSON OR PHOENIX, ARIZONA

a 'soldier boy,' but if we had been deputies they would have killed us all, never would they have been taken alive. The reward for capture was $7,000, all paid into the fund for Troop I of the 12th Cavalry."

The fugitives did not quibble over the legality of their capture in Mexico. The military issued a statement that they had pursued the men as draft dodgers, not murderers. Military officers refused to turn the men over to a posse until they signed a pledge to protect the prisoners with their lives. "That means," one soldier explained, "if you lose these prisoners to a lynch mob, you better be dead."

Brig Stewart had been appointed Graham County Sheriff to fill out McBride's term. Fearing he could not prevent a lynching, he sneaked the prisoners into Safford by way of Greenlee County.

An angry mob surrounded the courthouse in Safford. One by one, Stewart allowed them to file by the cells to view the prisoners. *The Graham County Guardian* reported that the captives, dirty and unshaven, appeared to be a "hard lot," and Sisson, an ex-convict, "looked right at home in his cell." Ragged, filthy, and suffering from their wounds, the prisoners endured jeers, threats, and the gut-wrenching fear that the crowd would become uncontrollable.

Tom and John contacted James Fielder, a lawyer in Deming, New Mexico, who had known their father. When Fielder's train left for Arizona, Charlie Power hopped aboard, stuffed $100 in the lawyer's hand and jumped off the train. Fielder put a quick stop to the viewing of the prisoners. He also waived the preliminary hearing and demanded a change of venue. Chambers, happy to oblige, agreed to move the trial to Greenlee County. A few years earlier, Greenlee had been part of Graham County, and Safford was still the trade center for both. That fact apparently escaped the out-of-state defense lawyer.

# 10

Clifton, the Greenlee County seat, lies in the copper-rich hills approximately thirty-five miles northeast of Safford. Nestled in a narrow, red rock canyon, the village spreads over the land like a colorful patchwork quilt. Brightly painted Mexican adobes, quaint Victorians, and nondescript shacks cling to the hillsides and line the San Francisco River. Heavy winter snows in the surrounding mountains bring spring floods to the canyons, and rocky bluffs radiate suffocating summer heat.

Until 1909 Clifton and the neighboring town of Morenci formed an industrial island in rural Graham County. Legislators represented agricultural interests in the county, but copper mines filled the state's coffers and fueled the economy. When mine owners demanded representation, eastern Graham County became Greenlee County. Safford maintained records for both until 1911, when the courthouse in Clifton opened for business.

Residents of Greenlee County traveled to Safford for goods and services unavailable at home. Many had relatives and friends in the Gila Valley. They heard the same gossip and read the same newspapers as the people who lived in Graham County. There was, however, one important difference between residents of Graham and Greenlee Counties. The latter depended on the mines for survival. From this pool of Greenlee County voters, lawyers would choose the jury for the Sisson/Power trial.

While the prisoners awaited their court date, hysteria over the war reached new heights. In Mexico, Pancho Villa raided an American-owned copper mine that supplied U.S. munitions factories. Twenty-six men died when a million-dollar power plant was blown to bits. In Chicago, the I.W.W. Union was on trial for seditious conspiracy. Banner headlines announced a union plot to overthrow the government. The I.W.W.'s war against war urged workers to defy the draft with insurrection, and many believed that was exactly what Sisson and the Power boys had done. They were "slackers" who had killed federal officers and, like many who were influenced by the I.W.W., they were also miners. Few seemed to know or care that Sisson, a veteran of a previous war, was too old for the draft. They failed to consider that Tom and John, self-employed mine owners, needed a union like a centipede needs a foot fungus. The prisoners lived in a world gone mad, in a time when men drew a thirty-year sentence for "unpatriotic utterances" and extremists demanded the death penalty for pacifists.

On Monday, May 13, 1918, the Sisson/Power trial opened in Greenlee County. In Clifton the weather was tolerable. Spectators, hoping to catch a glimpse of the defendants, swarmed outside the yellow-brick courthouse and crowded the inside halls. Upstairs, in the southwest corner of the building, deputies guarded escape routes from the courtroom and kept a wary eye on the prisoners. Everyone seemed to anticipate an eruption of violence before the trial ended.

Women pushed and shoved to get front row seats, and reporters were shocked by their number. Clustered together, the women whispered among themselves and craned their necks to eye the prisoners.

Seated at the counsel table, the defendants huddled around their attorney, James Fielder. John, obviously in pain, still wore a bandage over his left eye. Tom's wound had apparently healed, but his injured eye remained blind.

At the prosecutor's table, three county attorneys conferred in whispers: W. R. Chambers from Graham County, David Ling from Greenlee, and Norman Johnson from Gila County.[6]

The courtroom was suddenly silent. Everyone stood. Judge Frank B. Laine entered and took his seat at the bench. Built like an All-American tackle, his demeanor was solemn and stern.

Court reporter Ernest Shortridge scribbled "case number 438-A" on his notebook, and the trial began.

The State charged that Thomas Sisson, John Power, and Tom Power "feloniously, premeditatedly, and with malice and aforethought discharged a gun which did kill and murder Robert Franklin McBride."

The prosecution, of course, was not required to prove that all three defendants pulled the trigger of the same gun at the same time. Under Arizona law accomplices shared the guilt of the perpetrator who was, according to Chambers, Thomas Sisson.

The charge of murder in the first degree demanded proof of premeditation and malice. The State alleged:

- The defendants had lain in wait for the posse (premeditation).
- The first shot came from the cabin (malice).

[6] David Ling soon withdrew from the case and was replaced by A. R. Lynch.

- Jeff Power, a wanted man, resisted arrest.
- Sisson fired the shot that killed Frank McBride.

The defendants pleaded self-defense. Citing the time-honored tradition that a man's home is his castle, they claimed:

- The posse ambushed the cabin.
- Sisson did not participate in the shootout.
- Kane Wootan fired the first shot.
- Jeff Power, who had committed no crime, was shot while his hands were up by Deputy Wootan.

The tedious process of jury selection began. Tom and John, alert and interested, exchanged notes and occasionally whispered something to Fielder. Sisson seemed indifferent, resigned to his fate.

One by one, the lawyers questioned the panel. Prospective jurors intoned their names, addresses, occupations, etc. The day dragged on. Sun beat through the windows, and a great flutter of cardboard fans mingled with the drone of voices. Men mopped their foreheads with bandanas, and women tugged at their corsets, but nobody left the courtroom.

At five o'clock, court finally recessed for the day. On balmy May evenings, neighbors gathered around a pitcher of lemonade on somebody's porch or in somebody's yard. Those who had been in the courtroom reported to family and friends. They talked about the jury, and why so-and-so was selected while somebody else was not. They discussed the lawyers and described the defendants in detail. A few, touched by the youth of the Power boys, expressed regret that each had lost sight in one eye. Others voiced surprise at the mild-mannered demeanor of the defendants, but nobody questioned their guilt, not out loud.

The next morning when Tom entered the courtroom, he scanned the gallery, searching for his brother Charlie. Slumped into a chair next to John, he shook his head. Jury selection continued. Before noon, twelve men took their places in the jury box. With the possible exception of one rancher and one farmer, all were directly or indirectly dependent on the mines for their livelihood.

In the opening statements both the prosecution and the defense played on whatever prejudices they assumed the jurors had. The prosecution implied that the draft-dodging Power boys were somehow connected to the infamous I.W.W. The mere mention of the union conjured up images of industrial sabotage and violent overthrow of the government.

Fielder turned the argument around. He suggested that powerful mine owners had an interest in the conviction of the defendants. Most observers failed to follow his reasoning, but the odor of alcohol emanating from the defense counsel did not go unnoticed.

That night the prisoners tossed and turned on their cots. They still believed the posse had tried to kill them for their gold, but the futility of such an argument was apparent even to them.

The next morning Tom paused in the door to the courtroom and again searched the gallery for Charlie. A look of complete desolation crossed his face. Tom would never forget or forgive Charlie's absence.

Testimony in the trial began on Wednesday and ended at two o'clock on Thursday. Twenty-four names appeared on the witness list, but few were called to testify. The clerk's minutes detail the legal issues raised, but only newspaper accounts of the testimony remain. These accounts summarize the statements of only four witnesses: John Murdock, Deputy Marshal Haynes, Dr. Platt, and John Power. The defense tried to subpoena the man at the Redington Post Office who told Tom and John the army wanted only volunteers. Unfortunately, the army had shipped him to France.

At the inquest, Jay Murdock had testified that Tom Power said he'd punched out a window and shot Wootan. Jay's name was on the witness list for the trial, but he wasn't called. Jay's father, John Murdock, testified at the trial and told a similar story, but the names had been changed. John Murdock testified that Sisson said he punched out a window and shot McBride. If true, he was the only man at the camp who heard Sisson say he killed anybody.

On cross-examination, John Murdock admitted he'd been convicted of perjury in another case. More importantly, he was never sentenced. If Chambers chose to rectify that situation, John Murdock could spend considerable time in prison.

Both John Murdock and his son Jay witnessed Jeff Power's dying statement: "Why did Kane shoot me when I had my hands up beggin' for peace?" This declaration of a dying man was evidence vital to the defendants' case, but the jury never heard Jeff's words. The prosecution claimed the statement was inadmissible, and a firefight exploded in chambers.

The presumption that a dying declaration is truthful is based on the belief that a man who knows he's dying will not risk damnation by lying. Chambers claimed Jeff was a self-proclaimed nonbeliever, and Fielder was hard-pressed to deny the fact. Still, Jeff's mother had said, "Jeff *says* he don't believe in God, but he prays loud and long when he's sick." Obviously, in times of peril, Jeff never gambled on the nonexistence of God. Call it religion or superstition, the result was the same. Unfortunately, Judge Laine insisted on religion. Jeff's devotion to God, Laine decided, was not sufficient to ensure the truth of his statement.

The defense reeled from the blow. They were not aware that Jay had discovered Jeff's hands were up when the bullet entered his body, but Jay wasn't likely to rile the prosecution while a prison sentence hung over his father. Chambers' failure to produce Jeff's body at the inquest had destroyed any chance of finding evidence to support Jeff's dying declaration.

Dr. Platt took the stand and described the officers' wounds. He admitted he did not inspect Jeff Power's body, but no one cared. Some details in Platt's testimony had changed since the inquest. He no longer referred to the "heel mark" on Wootan's face as a bruise. Platt had decided the injury occurred after death. The implication that the corpse had been brutalized was even more damaging to the defendants.

Deputy Marshal Haynes was again the prosecution's star witness. More than three months had passed since the inquest, but Haynes' memory had improved. This time Haynes gave a name to the man he saw outside the cabin: Jeff Power. This time he remembered Jeff held a rifle across his chest. He even recalled that both Wootan and Kempton shouted, "Throw up your hands," at two-second intervals. Haynes said he saw Power drop the rifle to his right side, and he saw him fall, but he still insisted the first shot came from the cabin.

Once again, Haynes told the jury McBride carried *state warrants* for Jeff and Sisson. This time Chambers actually produced the warrants. When Fielder examined them, his face turned scarlet, and the veins bulged in his neck. Spewing fire,

Fielder turned to Chambers and, shaking an accusing finger at the prosecutor, he shouted, "The ink is still wet on these warrants!"

Justice of the Peace Paxton had issued the warrants without charges. Judge Laine declared them void. Chambers argued that officers were not required to judge the validity of a warrant. Void or not, he said, McBride had a right to act on the instructions in the documents.

Laine said officers are not qualified to make complicated legal judgments, but they are required to judge a warrant "on the face of it." In effect, Chambers argued the Sheriff was so ignorant of the law and lacking in common sense, he rode

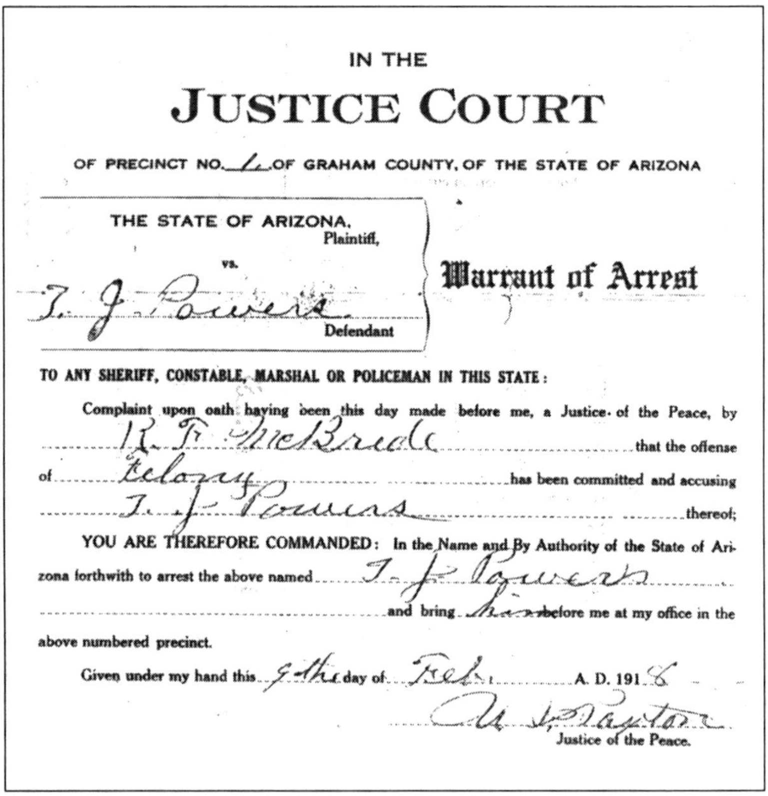

IN THE

JUSTICE COURT

OF PRECINCT NO. 6 OF GRAHAM COUNTY, OF THE STATE OF ARIZONA

THE STATE OF ARIZONA, Plaintiff,

vs.

T. J. Powers, Defendant

Warrant of Arrest

TO ANY SHERIFF, CONSTABLE, MARSHAL OR POLICEMAN IN THIS STATE:

Complaint upon oath having been this day made before me, a Justice of the Peace, by R. F. McBride that the offense of Felony has been committed and accusing T. J. Powers thereof;

YOU ARE THEREFORE COMMANDED: In the Name and By Authority of the State of Arizona forthwith to arrest the above named T. J. Powers and bring him before me at my office in the above numbered precinct.

Given under my hand this 9th day of Feb. A. D. 1918

W. S. Paxton
Justice of the Peace.

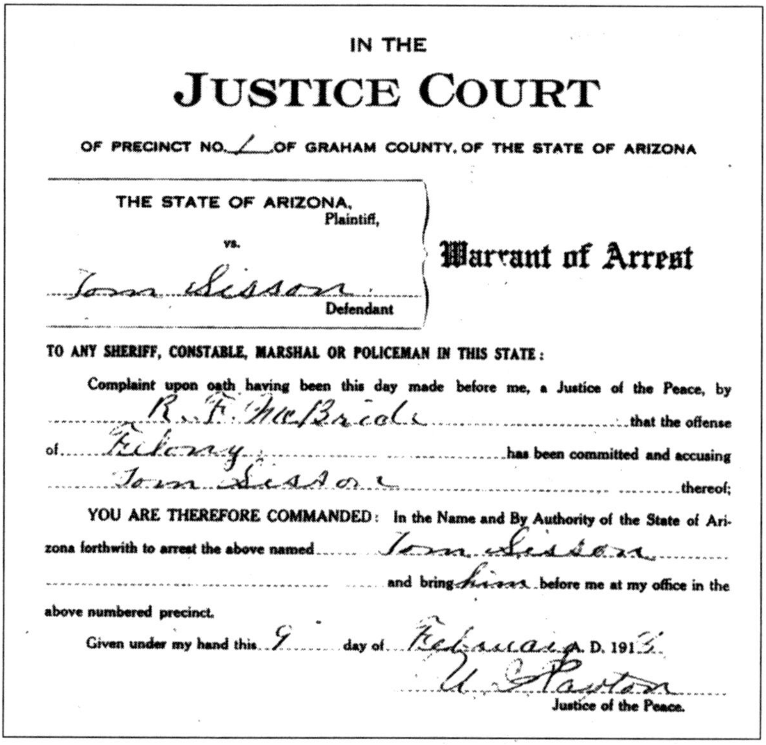
IN THE

JUSTICE COURT

OF PRECINCT NO. 1 OF GRAHAM COUNTY, OF THE STATE OF ARIZONA

THE STATE OF ARIZONA, Plaintiff,

vs.

Tom Sisson Defendant

**Warrant of Arrest**

TO ANY SHERIFF, CONSTABLE, MARSHAL OR POLICEMAN IN THIS STATE:

Complaint upon oath having been this day made before me, a Justice of the Peace, by R. F. McBride that the offense of Felony has been committed and accusing Tom Sisson thereof;

YOU ARE THEREFORE COMMANDED: In the Name and By Authority of the State of Arizona forthwith to arrest the above named Tom Sisson and bring him before me at my office in the above numbered precinct.

Given under my hand this 9 day of February A. D. 1918.

U. Lawton
Justice of the Peace.

twenty-five miles through a snowstorm in the middle of the night to arrest two men who were not charged with a crime.

The timing of the posse's arrival at the cabin suggested a surprise attack was planned, an ambush. Chambers said Tom Power's threat to kill McBride and his men after Ola's death was reason enough for a heavily armed posse to arrive at an unexpected hour. As for the officers' failure to identify themselves, Chambers declared there wasn't time before the shooting began.

Summing up the State's case, *The Tucson Citizen* reported that the prosecution proved Sisson killed McBride, but did not prove premeditation or malice. According to the newspaper, Chambers failed to prove the defendants knew the men outside were officers. "This knowledge is presumed," the reporter wrote, "but no evidence has been produced to this fact."

The evidence actually indicated the posse's arrival was completely unexpected. The cabin's glass windows, a clear danger to the men inside, had not been boarded or removed. Despite the Powers' barking dogs, the posse met no resistance as they descended the long trail behind the cabin.

The federal warrants for Tom and John were entered into evidence along with McBride's complaint. Like Marshal Dillon before him, Fielder questioned the right of a Justice of the Peace to issue federal warrants. Judge Laine ruled the warrants were valid.

John Power was the only defendant called to testify. He swore under oath that Kane Wootan fired the first shot and mortally wounded Jeff Power. John also declared Sisson never left his bed during the shootout.

*The Tucson Citizen* reported, "John proved to be a strong witness for the defense, minutely describing the tragedy. He said, after his father was shot down, he fired at both groups of officers, the fusillade was of a minute's duration, and, when firing ceased, he carried his father in." The article went on to describe the attitude in the courtroom. "Perfect order prevailed in all the proceedings. The indignation that existed during the manhunt is not apparent here and now."

After only a day and a half of testimony, the lawyers delivered their summations to the jury. Chambers concluded at three o'clock. At five o'clock the court recess interrupted Fielder's remarks, and he was forced to continue the next day.

Judge Laine's first instruction to the jury was a severe blow to the defense. Laine said, "Where a homicide is proven beyond a reasonable doubt, the presumption is the crime was murder in the second degree. If the State would elevate the crime to murder in the first degree, the prosecution must establish the characteristics of that crime. If the prisoner would reduce the crime to manslaughter, *the burden is on him*."

Laine's instruction violated a basic principle of criminal law. The accused has no burden of proof. The defendant is not even required to enter a plea. If the prosecution proves the accused killed the victim, but fails to prove premeditation and malice, the jury is obliged to return a verdict of manslaughter, excusable homicide, or justifiable homicide, according to the circumstances.

Judge Laine instructed the jury on the legal requirements for a verdict of first, second, or third degree murder, excusable homicide, and justifiable homicide. Premeditation and malice were necessary elements of first degree murder. Second degree required proof of malice, but not premeditation. The unlawful killing of a person without malice or premeditation constituted third degree murder or manslaughter. The use of weapons in this case precluded any consideration of excusable homicide, but justifiable homicide was a possibility. Homicide, Judge Laine explained, is justifiable when committed for the following reasons:

1. Resisting any attempt to murder or do great bodily injury to any person.
2. In defense of habitation, property, or person against one who intends to commit a felony by violence or surprise.
3. In the lawful defense of any person engaged in mortal combat, or in imminent danger, if that person has made a good faith attempt to decline further struggle.

4. When lawfully attempting to apprehend a person for a felony or in a lawful attempt to preserve peace.

As the trier of fact, Judge Laine explained, it was the jury's duty to determine the circumstances "as you believe they

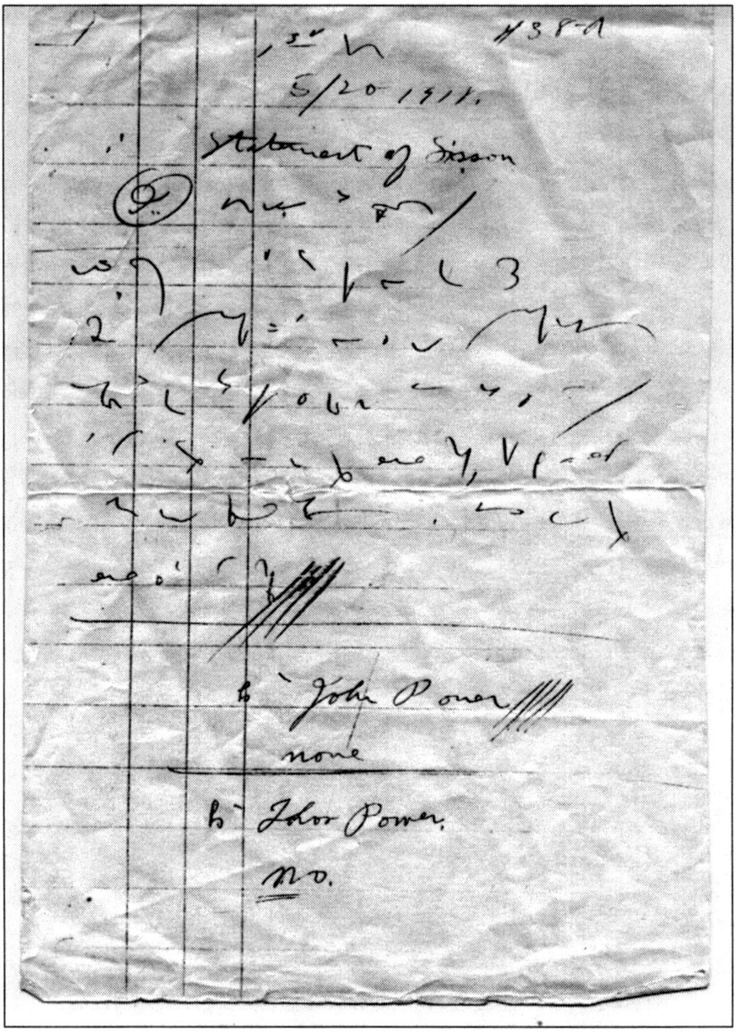


THE ONLY PAGE OF TRIAL TRANSCRIPT EVER FOUND

appeared to the defendants at the time, and not from any other standpoint."

According to *The Tucson Citizen*, the State produced no evidence of premeditation or malice. Nevertheless, the jury deliberated only twenty-five minutes before finding the defendants guilty of first degree murder.

On the following Monday, Judge Laine sentenced Tom Sisson, John Power, and Tom Power to life in prison, the maximum penalty under the law. Tom and John declined comment, but Sisson said, "I am not guilty of this crime. I plead with the court for mercy."

Judge Laine shook his head. "The court has no alternative in the matter. The jury has found you guilty in the first degree, and the law compels the court to pass sentence provided by statute. Consequently, I have no discretion in this matter."

This brief exchange between Sisson and Judge Laine was transcribed from the only page of the court reporter's notes ever found. Oddly enough, the docket shows Shortridge's notes were turned over to the court. Then Fielder filed notice of his intent to appeal. At this point, the court reporter had to retrieve his notes in order to transcribe them, but the notes, and any transcription he may have made, mysteriously disappeared. Without a record of the trial, an appeal was next to impossible.

Fortunately for Sisson and the Power brothers, capital punishment had been abolished in Arizona, but in the aftermath of the Sisson/Power trial, the death penalty was promptly reinstated.

Florence Prison, 1920

# 11

The Arizona State Prison at Florence swelled out of the desert like a festering wound. High walls topped by menacing gun towers surrounded the formidable institution. Built by convict labor from the infamous Territorial Prison at Yuma, the gloomy fortress seemed a monument to human misery.

Tom, John, and Sisson, stripped of freedom, property, and hope, entered the iron gates on May 22, 1918. The penitentiary, underfinanced and understaffed, had no clothes for new inmates. Soon reduced to wearing ragged shirts and pants, Tom, John, and Sisson went without coats in the winter. Meals were meager, tasteless, and monotonous. Prisoners ate hot cereal without milk or sugar for breakfast, stew for lunch, and beans or rice for supper. They worked five and a half days a week and returned to their cells at 6 PM for lock down. Night guards patrolled with flashlights, and smudge pots lit the main corridors. Tom and John, prisoners 5173 and 5174, were housed in cell twenty-two. The airless cell block reeked of the combined odors of kerosene, smoke, sweat, urine, and disinfectant. The recreation hall was equipped only with checkerboards, dominoes, and a few rickety tables and chairs. Tom and John spent most of their leisure hours in the library, poring over law books. The contrast between textbook law and law practiced in their own case infuriated them. They were entitled by law to a copy of the trial transcript, but every time John wrote the court, he either received no answer at all or they said there was no transcript.

Sisson, unable to read or write, relied on Tom and John to right the wrong of his conviction, but how could they help Sisson when they couldn't help themselves? They'd told everyone Sisson never left his bed during the gunfight. Over and over, they reminded them that the only evidence against Sisson was the word of a convicted perjurer, John Murdock. No one listened. No one cared.

Day and night, bitter memories of their father's death haunted the boys. Both the prosecution and defense agreed that the posse had not identified themselves. Even Haynes admitted their father was the first man shot. Why couldn't the jury see they had acted in self-defense? Tom, enraged by what he considered Chambers' arrogance, audacity, and injustice, vented his anger on anyone willing to listen. The Prosecutor, he complained, hadn't even bothered to hold an inquest into Jeff's death. Without a shred of evidence, they had accused Jeff and his sons of murdering their beloved Ola, and they'd mutilated her body trying to prove it.

As alone as they felt, the prisoners were not without friends. Charlie Power wrote to his brothers often. John answered, but Tom couldn't forgive Charlie for his absence at the trial. The Morgans, Habys, Claridges, and other ranchers in the Aravaipa did everything they could to help, but theirs was the proverbial voice in the wilderness.

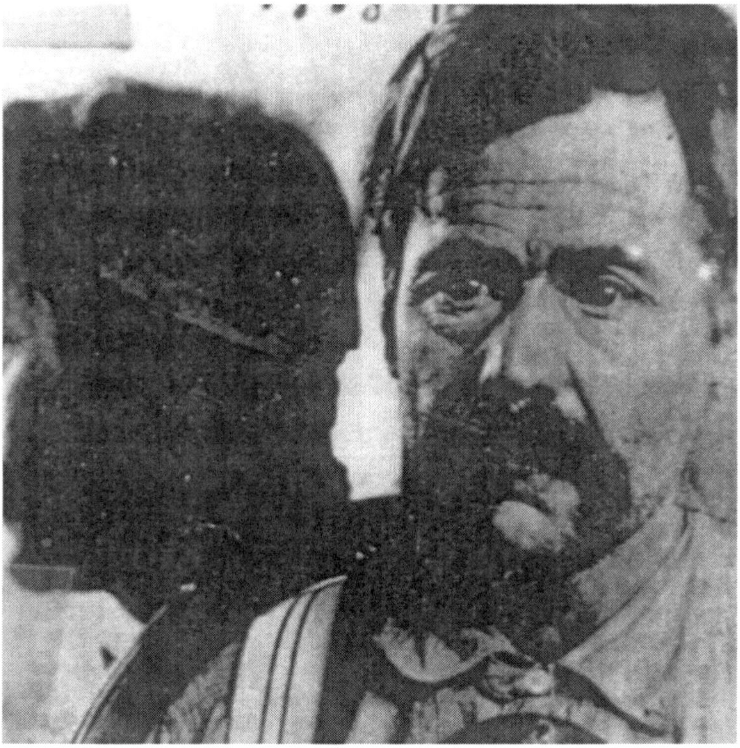

Tom Sisson

The conviction of the Power boys and Sisson left Graham County bitterly divided. Few in the valley knew the prisoners, but many were related by blood or marriage to the Kemptons and McBrides. They saw the widows struggle to raise their children alone. They saw the deprivation some of those children endured. A jury had convicted the prisoners of murder, and that was good enough for them.

Contrary to popular belief, the battle lines were not drawn strictly between Mormons and non-Mormons. Some ranchers, who fought long and hard for the prisoners, belonged to the church. A few Mormons in the valley were not unaware of legal irregularities in the case. According to rumor, the Powers were guilty of everything from incest to murder, yet some observers remained skeptical. They discreetly questioned people who actually knew the Power family. A merchant who had done business with the Powers for years looked over both shoulders before he replied, "Jeff Power was the nicest man you'd ever want to meet, and honest as the day is long." Gregory Haby was less reticent. He described the family as "Good, hard-working people." Haby declared, in no uncertain terms, "Jeff Power's word was as good as his bond." These proclamations had little effect on majority opinion, however. Mormon clannishness

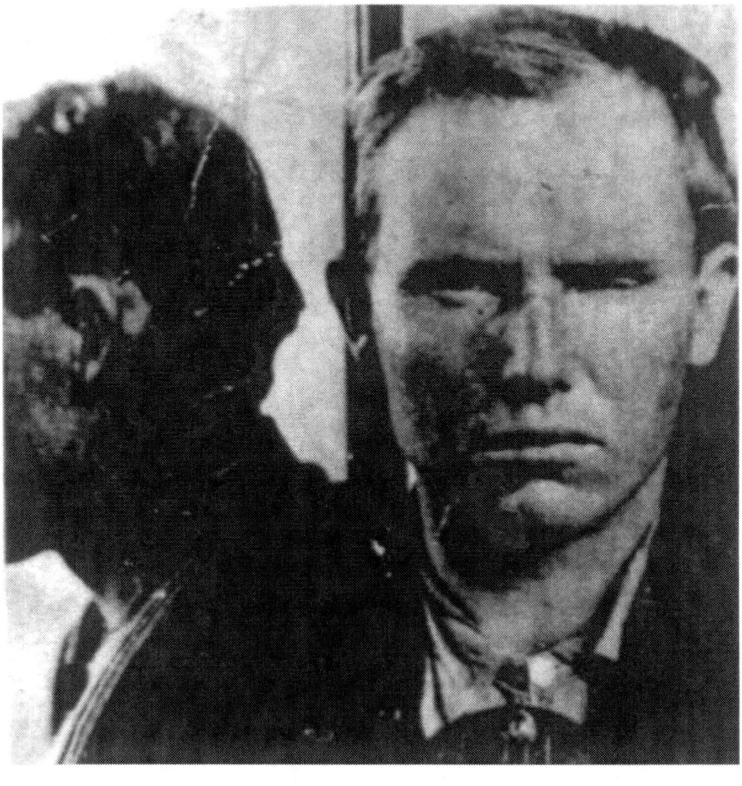

JOHN POWER

TOM POWER

discouraged expression of divergent views. Most who questioned the guilt of the prisoners found it more politic to keep their opinions to themselves.

On August 14 the widows' suit against the Powers for the wrongful death of the officers came to trial in Graham County. Tom and John could not attend, of course. Laura Wootan, Sena Kempton, and Clara McBride won a $20,000 default judgment against them. The widows took over the Powers' mining claims. According to rumor, the claims were worthless, but Clara McBride paid $2,500 for a portion of Sena Kempton's share.

The Power boys sat helpless in their cells while everything they owned was sold to satisfy the judgment against them. The sheriff auctioned machinery, ore cars, rail track, mining tools, blacksmith equipment, furniture, livestock, saddles, guns, traps, kitchen utensils, Jeff's car, Ola's violin, Granny's sewing machine, sugar, flour, and even a two-pound tin of baking soda. Clara McBride bought most of the mining equipment.

The prisoners, feeling hopeless and helpless, continued to fight for justice. John wrote more than twenty-seven letters pleading for a hearing before the parole board, but the group in Safford lobbied against them. More often than not, John received no response to his letters.

One day followed another. August's sweltering heat became September's sweltering heat. Work was the only outlet for the prisoners' misery and frustration. Sisson and the boys, accustomed to hard labor, threw themselves into any job assigned to them. Prison officials appreciated their industry. As time passed, they achieved trustee status. All three lived and worked outside the prison walls: on the prison farm, in the prison dairy, and in the kennels. In the summer, when school was out and the guards' children were underfoot or into mischief, their fathers sent them over to help Tom and John with chores. The Power brothers enjoyed the youngsters and taught them to milk cows, train dogs, and ride horses.

Sisson, a veteran of the Indian Wars, received a monthly stipend from the government and saved every dime. Time and again, the three inmates saved enough money to hire a lawyer. Several attorneys accepted a fee, but due to the lack of a trial transcript, nothing ever came of their efforts.

"Lifers" got no time off for good behavior, and Tom had nothing to lose when he walked away from the prison in 1923.

Tom and another inmate enjoyed a brief spree in San Diego before police returned them to Florence a month later.

In 1926 two lawyers petitioned for a pardon on behalf of Tom, John, and Sisson. Former County Attorney Chambers was now Judge Chambers of the Graham County Superior Court, and he responded with blazing guns. He wrote a letter to the Arizona Board of Pardons and Paroles, protesting the release of the Power brothers and Sisson. Chambers' letter said:

> "A copy of the petition of Tom and John Power, convicted in the Superior Court of Greenlee County in May 1918 for the murder of Robert F. McBride, Sheriff of Graham County, has come into my hands. As I have told friends and relatives of Sheriff McBride and his deputies, Martin Kempton and T. K. Wootan, who were killed at the same time, I do not anticipate that the present board will take any favorable action upon this petition, but the ravages of time may take many of us away, so I desire now to enter my protest against the pardon of these men, not only for today, but for all time.
>
> "When this murder was committed, I was County Attorney of Graham County. Assisted by Norman J. Johnson of Globe and A. R. Lynch, then of Clifton, I conducted the prosecution against John and Tom Power and Tom Sisson. They were defended by James S. Fielder of Deming, New Mexico, in his palmy days rated as one of the best criminal lawyers of the Southwest.[7] Norman

[7] Fielder's "palmy days" were apparently over when he defended the Power boys and Sisson. Judge Laine said the defendants had not had adequate counsel at the trial.

Johnson and James S. Fielder have already passed over the divide. As County Attorney, I made a personal investigation of every scrap of evidence introduced upon trial, and am probably more familiar with the case than any other living man. That evidence, so far as I know, was never transcribed. At the conclusion of the trial and judgment and conviction, Mr. Fielder, attorney for the defendants, did not think enough of the case to even appeal it, when he knew he could do so at the State's expense.[8]

"After the closing of the State's case, the principle evidence offered on behalf of the defendants was the testimony of the defendant, John Power. He made such a sorry spectacle of himself that Mr. Fielder would not even put Tom Power or Sisson on the witness stand.

"The case is entirely too long for me to enter into an exhaustive review of the evidence. Suffice to say that the conviction is supported by the testimony of the only eye witness, aside from the defendants, Frank Haynes, then Deputy U.S. Marshal, by admission of the defendants after the murder, and a long list of threats covering a period of several months.

"The court reporter who reported the case is still alive and available to the defendants and the State. The testimony should be transcribed, not only for the use of the present Board of Pardons and Parole, but for the use of future boards.[9] This sworn testimony is much more reliable than the statements of two lawyers who

---

[8] Mr. Fielder filed notice of his intent to appeal, but could not appeal without the missing transcript.

[9] The court reporter's notes have never been found.

were not present at the trial and know nothing about the facts. The oath of the defendants, John and Tom Power, attached to the petition, should carry no weight with a board of sane and intelligent men. In making a garbled statement of fact in connection with the petition, which, to my knowledge is mostly falsehood, this petition standing alone might lead someone to believe a gross injustice had been done.

"On the contrary, these defendants were granted a fair and impartial trial, and the verdict was not the result of passion or prejudice. As County Attorney, feeling that the defendants might not get a fair trial in Graham County, and to avoid all danger of a lynching, I consented to a transfer of the case to Greenlee County, where we had a good jury drawn in the regular manner where about eighty percent of the citizens live in mining camps.

"Any pardon of these men would be an outrage upon justice and would turn loose upon society two of the most desperate criminals ever convicted of murder in this State."

Chambers' gross misrepresentation of fact should have been apparent to anyone who bothered to look at the few scraps of paper in the court file. Marshal Dillon's letter of instruction to Haynes was in the file, as was Haynes' perjured testimony regarding the instructions. The warrants for Jeff and Sisson, which contained no charges, were also there in the record. If that wasn't enough to raise questions in a sane man's mind, there was the mysterious disappearance of the court reporter's notes. That fact alone flashed a warning light bright enough to yellow the mind of a first-year law student, but the board denied the Power brothers' petition for pardon, presumably on the strength of Chambers' letter alone.

# 12

DEVASTATED BY THE BOARD'S REFUSAL, it seemed to Tom and John that some cruel, preordained destiny was their lot in life, and nothing and no one could change it. Pessimism overwhelmed them.

Although most "lifers" served less than ten years in prison, Sisson and the Power boys remained through the great depression. By 1939 Sisson was well into his seventies, and Tom and John were pushing fifty. They had served more than twenty years without a hearing before the Board of Pardons and Paroles. It was a Catch-22. Without a transcript, they had no trial record on which to base an appeal. A parole or pardon could be argued, perhaps, on the basis of their conduct in prison. They were trustees, the guards trusted them with their children, their record was good, but Chambers' letter had destroyed any possibility of pardon or parole.

As the year drew to a close, Tom Power thought he saw Braz Wootan patrolling the prison wall. According to Tom, the warden denied he had hired Wootan. Tom and John suspected a "setup" and ran for their lives. Later, Tom, with his usual wry humor, explained the situation to a friend. "I says, 'Warden, this prison ain't big enough for me and Braz Wootan. One of us has gotta go.' Braz stayed, so me and John went."

Tom and John fled to Mexico. Aided by friendly Mexicans, the brothers rented an apartment in Mexico City and toured the country. Without papers, neither could find work. After three and a half months, their pockets were empty. Crossing the Rio Grande into Texas, their luck, like their money, ran out. Texas officers had set a trap for a man who had brutally murdered his wife and child in Mexico. The Power family curse raised its ugly head, and the trap sprang shut on Tom and John. The astonished officers recognized the Power brothers and sent them back to Florence.

In the late thirties, another war loomed on the horizon. German troops marched into Austria and Poland, and history repeated itself. As most of Europe fell under Hitler's domination, complacent Americans ignored the warning. Echoing Jeff Power's alleged sentiments of two decades earlier, they dubbed the conflict "Europe's war."

At dawn on December 7, 1941, Japanese planes bombed Pearl Harbor. Four days later, the United States plunged into a bloody war with Germany, Italy, and Japan. As in World War I, fear and suspicion flourished. American citizens of Japanese ancestry were forced into internment camps. This was no time to advance the cause of two convicted draft dodgers from a previous war. It seemed the family curse was always there lurking in some dark corner. Even Charlie Power could not escape the dreadful consequences. All four of his children died in World War II.

The year 1945 brought an end to the mightiest struggle mankind had ever seen. In the decade that followed, relative peace restored sanity to a rapidly changing world. Sometime between 1952 and 1955, the Parole Board granted a hearing to Tom and John Power. Several County Sheriffs and former Prison Wardens supported their release. Sisson's prison record was spotless, but he was not considered for parole because he was too old to work. As a veteran of the Indian Wars, he received a pension of $100 a month. He had $10,000 in savings, and he was entitled to health care and lodging in a home for old soldiers. He didn't need to work. The State's paternalistic attitude was cruelly patronizing and unjust, but Sisson had no choice but to accept the decision.

At the hearing for Tom and John, Graham County Attorney W. R. Richardson defended the interests of the slain officers' families who gathered en masse for the hearing. Like so many before him, Richardson had not bothered to read the court record. Protesting the release of the prisoners, he parroted old tales of the Powers' "fortress" in the mountains and their "barricaded" cabin. The inquest transcript he'd supposedly read disproved his statements, but Richardson relied on old rumors he apparently believed.

Widows, children, and grandchildren of the slain officers applauded Richardson's remarks. Some were not yet born when the Powers entered prison. Tom and John had no history of violence before or after the shootout, yet relatives of the officers claimed to fear for their lives if the men were released.

That so much bitterness had survived more than thirty years astounded prison officials. More appalling was the families' unshakable faith in groundless rumors denied by evidence. Like the bitter hatred that blinded them, the rumors had become a family legacy passed from generation to generation. Most of the protesters were good people by any standard, but their self-righteous loathing of the Powers was shocking in its magnitude and horrifying in its intensity.

No board of political appointees was ever likely to stand against such pressure from the Mormon communities within the state. The board denied the Powers' request for clemency.

# 13

In the years that followed the clemency hearing, Sisson traveled around the prison yard with the aid of a cane or crutches. In January 1957, the old scout entered the prison hospital. On the twenty-third, he died. According to Army records, Sisson was eighty-nine. Not one single infraction of prison rules appeared on his record, and that record spanned forty years, counting the time he served for horse theft. Sisson had saved $10,000 from his army pension, and he left the money to John. In death, Sisson was accorded the dignity denied him in life. Tom and John refused to bury their old friend in the prison graveyard. John went to probate court to claim his inheritance, and he bought a coffin and a plot in the Florence cemetery. John and Tom buried Sisson with the respect and affection he deserved.

Sisson's death marked the end of an era for the Power boys. Later that year, Russia launched the first satellite into space. Tom and John, who had only a nodding acquaintance with the automobile, seemed forever lost in a time warp.

The following year, a chance meeting at the prison between Warden Frank Eyman and journalist Don Dedera was serendipitous for the Power boys. Dedera, a popular columnist for *The Arizona Republic*, the state's largest newspaper, had come to the prison to interview a particular inmate. Eyman suggested the Power brothers' plight would make a better story. Dedera was not yet born when Tom and John went to prison, and he knew next to nothing about the case, but Eyman was persuasive. Dedera met with the brothers and promised to look into their situation.

The reporter studied old newspapers, read the inquest transcript, and questioned a few people who remembered the case. He knew the odds against a commutation of sentence were enormous, but he accepted the challenge. The prisoners were mild-mannered old men, and after forty-one years of confinement he believed they deserved a break.

For nearly two years, Dedera argued in his column for the release of the Power boys. Protests poured in from Graham County, Mesa, and other Mormon strongholds in the state. Rumors from forty years past were cited as fact. Again and again, letters claimed the prisoners ambushed the posse. Protesters claimed Kane Wootan was shot while still on his horse. They insisted the Power boys shot McBride in the head while he lay wounded and helpless on the ground. One writer proclaimed the posse had *legal* warrants for everyone in the cabin, and they all believed the prisoners had a fair and impartial trial.

Dedera answered every letter. Eventually, old rumors were replaced by professed concern for the Powers' welfare. Protesters claimed the prisoners had remained in prison too long and couldn't take care of themselves in a modern world. In prison, they had a good home, good food, and good medical care.

Dedera faced the issue head-on. In an award-winning column, he asked if it was the policy of the Mormon Church to insist the Power boys die in prison. The question marked the turning point. Church leaders strongly denied that such a policy ever existed, and some prominent members spoke out in favor of the Powers' release.

On April 20, 1960, due to the persistent efforts of Don Dedera and *The Arizona Republic*, the Power boys appeared before the Arizona Board of Pardons and Paroles. Relatives of the slain officers packed the room. Clara McBride had agreed not to protest the Powers' release, but the Kemptons bypassed Graham County Attorney Alvin Krupp and hired W. R. Richardson to defend their interests. Richardson, who still hadn't reviewed the Greenlee County records, insisted that the widows would not sleep nights if the Power boys were released. He reminded the board that Tom and John had shown no remorse for their crime.

Old friends of the Powers spoke on their behalf, and former Warden Lorenzo Wright made an impassioned plea. According to *The Arizona Daily Star*, Wright, former Stake President of the Church of Jesus Christ of Latter-Day Saints, said he knew the Wootan, Kempton, and McBride families. "I loved them all," he said. "Some wonder why I am here. I have no personal interest in the Power boys, but I feel it would be contrary to my beliefs and trust if I did not come here and plead for my belief. I am a churchman. I feel I represent God Himself today. Of course, I hold high priesthood. God has said if we do not forgive, he will not forgive us. They (the Power boys) have to face Him and final judgment. They will have a lot of repenting to do, and will not be able to hold a grievance. I am only here for the betterment of man. I am here on earth to help the underdog to meet his Maker."

Then Wright turned to Tom and John and demanded, "Will you forgive and ask forgiveness?"

Tom flushed. Slowly, quietly, he said, "We've got nothing against anybody. We're not acquainted with any of the families. We didn't know any of the men except Wootan. It was out of our control—what happened forty-two years ago. We're sorry."

All eyes turned to John. Tom glanced at his brother and shoved his elbow into John's ribs. John rose slowly. Choking on every syllable, he said the words. "We're sorry...we beg forgiveness." The whole room applauded.

The board granted parole. Old friends and prison officials swarmed around the stunned prisoners. For a moment, John and Tom seemed unable to grasp the fact they were free. They had served the longest term on record at the Arizona State Prison—forty-two long years—and they could not hold back their tears.

A week after the hearing, John and Tom left the prison with new Levis, boots, hats, and $12.50 in mustering-out pay. Tom was 68, and John had turned 70, but there was still time for a couple of old cowboys to wrangle a steer or two.

During the decade that followed, John and Tom tried to make sense out of the changing world outside of prison. A half century had passed since they failed to register for the draft. That mistake had destroyed their lives. Now America was again at war. A new generation rioted in the streets, burned their draft cards, and torched the stars and stripes with impunity.

John returned to the peace of the Aravaipa, where little had changed. Many old friends had died, but their children and grandchildren remained.

Tom, like his father before him, seemed afflicted with wanderlust. He made several visits to California, but he always returned to the life he knew best.

The Power boys were grateful for their release, but not satisfied with parole. For ten years, they fought for a full pardon.

When the State of Arizona finally restored their citizenship rights, they voted for the first time. In 1916, when Kane Wootan barred them from registering, they had planned to support the Democratic ticket, but in 1970 they registered as Republicans.

Today, on a mesa overlooking the Aravaipa, lies all that remains of the star-crossed Power family. Tom and John's final words on the shootout are carved in stone on their father's grave: "T. J. Power, Sr. shot down with his hands up in his own door."

The arrangement of the graves is interesting. There was room for four in the original plot. Jeff's mother and Ola are buried there side by side. John retrieved his father's remains from a shallow grave near the mine, but he didn't bury him in the original plot. John bought a second plot big enough for two next to the original. When Tom died, John buried him next to their father. The two spaces left in the original plot, John saved for his mother and himself. Unfortunately, Mattie

Morgan Power had been buried in New Mexico too long, and there was nothing left of her to move. John was the last to die. Crippled with arthritis and neuritis, he could barely walk, but he dug his own grave in the original plot. Some thought it odd that John chose to be buried with the women. "Tough as nails he was," they said. But John was the sentimental son, and perhaps he dreamed of kinder, gentler companions in death than he'd known in life. The inscription on his headstone speaks of his final hope: "John Grant Power rest in peace."

# Afterword

In 1985 Dr. Richard Froede and Dr. Allen Jones studied the facts and rumors surrounding the death of Ola Mae Power. At that time, Dr. Froede was Pima County's Chief Medical Examiner and Professor of Forensic Pathology at the University of Arizona. Dr. Jones, also a Pima County Medical Examiner, was Professor of Clinical Pathology at the same university.

Years earlier, during his autopsy, Dr. Platt had noted a dislocated cervical vertebra on Ola's body. Drs. Froede and Jones believe the injury was actually a fracture and occurred postmortem. Dr. Froede explained, "A cervical fracture is sometimes impossible to detect with the naked eye, even when the spine is removed and examined by a physician. The neck of an adult cadaver is extremely vulnerable to injury. Dr. Jones and I handle bodies every day. We are trained, experienced, and very careful, but we sometimes damage a neck."

The doctors explained away many popular misconceptions regarding the timing of Ola's death. The rumor that the slippage of Ola's skin indicated she'd been dead for days was erroneous. Dr. Jones said, "Skin slippage can occur minutes after death. It simply indicates circulation to the skin has ceased."

Both doctors agreed the stench in the cabin, the extent of rigor mortis, and the discoloration of the body were all signs of normal decomposition consistent with the time of death reported by the Powers.

The significance of Ola's dying word, "Poison," had inspired more erroneous rumors. Drs. Froede and Jones ruled out strychnine, arsenic, and other common poisons as the cause of Ola's death. Her symptoms, they said, were not consistent with these toxins. In all probability, they agreed, Ola Mae Power died of botulism.

Cynthia Ratner, food processing expert and acting Director of the Agricultural Extension Service in Tucson, reached the same conclusion.

"In the early 1900s," she said, "canning was a primitive art compared with today's methods." According to Ratner, the time, place, and circumstances of the Powers' existence guaranteed the presence of botulism in some of their food. "Botulism bacteria grows in the absence of oxygen, and canned food is fertile ground," Ratner explained.

At the Power's home in Rattlesnake Springs, "Power's Garden Place," Ola and Granny had canned enough food to last for years. From that home, they moved down to Klondyke, up to Rattlesnake Canyon and up again to Kielberg Canyon. Their food supply traveled with them. Jars are jostled in transit, and seals are unavoidably broken. Under these circumstances, according to Ratner, contamination was inevitable.

Dr. Froede suspected the sauerkraut Ola had prepared for dinner was the contaminant.

Ratner said, "Botulism does not usually occur in high acid foods, but the degree of acidity depends on the method of preparation. When foods are prepared for canning, they must reach a certain temperature to kill bacteria. That degree of heat must be maintained for a period of time that varies with altitude. Some recipes called for vinegar. Most people made their own, and the degree of acidity was unpredictable.

"They didn't know much about botulism in those days." Ratner sighed. "A lot of people died, and nobody knew why."

With her dying word, "Poison," Ola had warned her family away from the food she'd prepared. She'd saved the lives of her brothers, her father, and Sisson. How ironic that baseless rumors and suspicion surrounding Ola's death led to their destruction.

## About the Author

Barbara Brooks Wolfe is a graduate of the University of Arizona in Tucson and a native of the State. A former teacher, she is an avid history buff and researcher.

## In Appreciation

The publisher and author wish to express their thanks to *The Tucson Citizen* for permission to reprint articles relating to the Power case, and especially to Meggin Kitterman, who was extremely helpful in providing copies of the originals.

Our thanks also go to Terry Ridgeway for permission to reprint photographs from the collection of his father, historian W. R. Ridgeway of Safford, Arizona.

Printed in the United States
127757LV00002B/11/P

9 780979 934179